FIBEROPTIC BRONCHOSCOPY

FIBEROPTIC BRONCHOSCOPY

in Diagnosis and Management

Editors:

Roland M. du Bois MA MD MRCP
Consultant Physician and Senior Lecturer

Stewart W. Clarke MD FRCP FCCP
Consultant Physician and Senior Lecturer

Royal Free Hospital and School of Medicine
London, UK

Foreword by:

Sir John C. Batten KCVO MD FRCP
Honorary Physician to the
    Brompton and St. George's Hospitals
London, UK

Grune & Stratton, Inc.
Harcourt Brace Jovanovich, Publishers
Orlando   New York   San Diego   London
San Francisco   Tokyo   Sydney   Toronto

Gower Medical Publishing   London · New York

Distributed in the USA and Canada by:
Grune & Stratton, Inc.
Orlando, FL 32887
USA

Distributed in all countries except the USA, Canada and Japan by:
J.B. Lippincott Company
East Washington Square
Philadelphia, PA. 19105
USA

Distributed in Japan by:
Nankodo Company Limited
42-6, Hongo 3-Chome
Bunkyo, Tokyo 113
Japan

Project Editors: David Goodfellow

Melanie Paton

Design: Michael Laake

Illustration: Chris Furey

Printed in Singapore by Imago Productions (FE) PTE Limited.
Originated in Hong Kong by South Sea International Press Ltd.
Typesetting in Bournemouth by Ampersand Typesetting Ltd.

Text set in Garamond ITC book and Garamond ITC book condensed; captions and artworks set in Univers.

British Library Cataloguing in Publication Data:
Fibreoptic bronchoscopy: in diagnosis and management.
    1. Bronchoscope and bronchoscopy   2. Fibreoptics
    I. du Bois, R. M.   II. Clarke, Stewart W.
    616.2'307545   RC734.B7

ISBN: 0-8089-1881-8

© Copyright 1987 by Gower Medical Publishing Ltd., 34-42 Cleveland Street, London W1P 5FB, England.

Foreword

When I first came to the Brompton Hospital in 1950 we used to go to the operating theatre to see our patients bronchoscoped by the surgeon through a rigid instrument. This was not only an educative process for us all but enabled us to discuss, where appropriate, the case with the surgeon. Fear of cross-infection then brought about a bleak period for physicians when they were for practical purposes excluded from the theatre, and bronchoscopy was the reserve of the surgeon.

Fibreoptic bronchoscopy changed all that and Stewart Clarke, who came to the Brompton Hospital in 1971, pioneered the use of this instrument in the UK. Within a year or so he had broken loose from the operating theatre and was performing fibreoptic bronchoscopy under local anaesthesia. Since then fibreoptic bronchoscopy has become the perquisite of the thoracic physician, is part of his training and is now a major diagnostic and therapeutic tool.

Stewart Clarke reminds us in his introduction how fibreoptics were discovered and then the fibrescope itself in 1954 when Hopkins and Kepany demonstrated the transmission of coherent images through bundles of aligned glass fibres. Technological development then passed from the UK to North America but particularly Japan. Stewart Clarke and his colleagues using the refined flexible bronchoscope illustrate clearly the diagnostic and therapeutic potential of this remarkable instrument, including the latest developments in laser therapy and in physiological measurement. The publishers are to be commended for the clarity of the text and the diagrams and for the radiographs and illustrations of the highest quality.

I can recommend without reservation this excellent manual to all thoracic physicians and surgeons, not only those who are learning to use this instrument, but also to those who are well versed in the technique.

J.C.B.

Preface

Since the fibreoptic bronchoscope was introduced into this country in the early 1970's, the numbers using it and the range of uses to which the instrument can be put have increased enormously. Samples obtained by fibreoptic bronchoscopy are now being analysed in greater detail than ever before and this makes the instrument a powerful tool in the hands of the trained observer.

It is the aim of this volume to provide a visual manual of fibreoptic bronchoscopy, principally for those in training. The techniques of bronchoscopy and procedures that can be performed through it, investigation of lung problems, physiological aspects of bronchoscopy and therapeutic uses of fibreoptic bronchoscopy have all been covered. Throughout, the emphasis has been on a visual presentation of practical aspects in the management of patients with pulmonary disease and for this reason the basis of each chapter has been the presentation of individual problems.

It is hoped that this book will be of interest to those who wish to learn fibreoptic bronchoscopy and also those who might like a review of the subject.

A large number of people have contributed to the production of this volume. We are grateful to Dr P. Corris, Dr P. Griffiths, Dr C. Grubb, Dr R. K. Knight, Dr A. Newman-Taylor and Prof. M. Turner-Warwick for their help with some of the photographic material. We are particularly indebted to David Goodfellow for his enthusiasm, diplomacy and patience.

R. du B. S.W.C.

Contributors

Martin R. Hetzel MD FRCP
Consultant Physician, Whittington Hospital, Consultant
Thoracic Physician, University College Hospital and Honorary
Senior Lecturer in Medicine, Faculty of Clinical Sciences,
University College, London, UK.

Michael D. L. Morgan MD MRCP
Senior Registrar, Department of Respiratory Medicine, East
Birmingham Hospital, Birmingham, UK.

David E. Stableforth MA MB BCh FRCP
Consultant Physician, Department of Thoracic Medicine, East
Birmingham Hospital and Honorary Senior Lecturer, University
of Birmingham, Birmingham, UK.

Acknowledgements

Chapter 1
Figure 1.1a provided by Key Med.

Chapter 3
Figures 3.17, 3.18, 3.24 and 3.35 reproduced from Stradling P,
*Endobronchial Anatomy and Pathology. 100 Colour
Slides for Medical Teachers.* By courtesy of the publishers,
Boehringer Ingelheim.

   The endobronchial photographs were taken with an Olympus
OM1 35mm camera and CLV10 light source, kindly loaned by
Key Med, unless otherwise stated in the captions. Figure 3.5a
provided by Key Med.

Chapter 4
Figure 4.20 reproduced from Davison AG, Haslam PL, Corrin B,
Coutts II, Dewar A, Riding WD, Studdy PR & Newman-Taylor AJ
(1983). Interstitial lung disease and asthma in hard-metal
workers: bronchoalveolar lavage, ultrastructural and analytical
findings and results of bronchial provocation tests. *Thorax,* **38,**
119–128. By courtesy of the publishers, *Thorax.*

   Grateful thanks to Dr S. Milkins, Department of
Histopathology, Royal Free Hospital, London, UK, for
preparation of many of the photomicrographs.

Chapter 5
Most of the physiological tests described were adapted for
the fibreoptic bronchoscope by Professor D. M. Denison.
Ms C. Busst is thanked for her technical assistance.

Chapter 6
The anaesthetic protocol shown in Figure 6.18 was developed
by Drs C. Nixon and C. Childs, Department of Anaesthesia,
University College Hospital.

   I would like to acknowledge the support and technical
assistance of Sister A. Ramu, Dr S. Bown and Dr T. Mills in the
National Clinical Laser Unit, University College Hospital,
London, UK.

Contents

Introduction

It is often said that there is nothing new under the sun. With bronchoscopy, Hippocrates suggested intubation of the larynx and trachea in cases of asphyxia over two thousand years ago. Furthermore, instruments for inspecting the various bodily orifices were found in the ruins of Pompeii. Progress with bronchoscopy was remarkably slow from then until the nineteenth century when Horace Green (1846) catheterized the trachea and bronchi for the first time. He presented his findings to the Medical and Surgical Society of New York in 1847, when his technique was condemned by the members as 'an anatomical impossibility and an unwarrantable invasion in practical medicine'; he was promptly expelled.

By contrast, a singing teacher named Manuel Garcia fared rather better. Not unnaturally he was particularly interested in the function of the vocal cords and when he succeeded in observing his own in action, by means of mirrors, he wrote a paper in 1855 entitled *Physiological Observations on the Human Voice*. He was appointed Professor at the Royal Academy of Music in London, had the distinction of training Jenny Lind and thereafter presenting his paper to the Royal Society, whereupon he was elected a Fellow.

The removal of inhaled foreign bodies remained a recurrent stimulus to the further development of bronchoscopy. In the literature, many unusual cases of choking are described and in two particularly celebrated cases the patients choked while attempting to swallow live fish which became impacted in the larynx. The prevention of this is obvious though the treatment then was less so.

It is generally agreed that Gustav Killian of Freiburg was the first to remove a foreign body lodged in the bronchial tree, by means of a rigid bronchoscope in 1897. He practised assiduously on cadavers beforehand and made many adaptations both to the equipment and the technique, including the use of his floating bronchoscope which enabled the operator to have both hands free for manipulation. His techniques were subsequently taken up by Chevalier Jackson (1907) in the USA and developed to a fine art. These pioneers used a simple rigid tube which has changed little since its inception, although lighting, telescopes, ventilation and anaesthesia have all been improved.

The flexible fibreoptic bronchoscope is a product of the last two decades. Its origin, however, which was mainly in Britain, can be traced back to Tyndall in 1870. He noted that the fine silk-like threads produced by pulling apart a melting glass rod had well defined optical properties, such that light could be trapped and transmitted around bends by internal reflection. Later, Baird, in 1927, explored this property further and filed a patent for an image-carrying curved glass tube (British patent specification No. 20, 969/27). Thereafter he devoted himself to the development of television, in which again he was a pioneer.

Further progress was sluggish until in 1950 Van Heel found a means of improving the efficiency of internal light reflection by coating each fibre with a glass layer of lower refractive index. In 1954, Hopkins and Kapany summarized their discoveries on transmission of coherent images through aligned bundles of specially coated flexible glass fibres and coined the term 'fibrescope'. With a good deal of foresight they commented: 'an obvious use of the unit is to replace the train of lenses employed in conventional endoscopes'. This suggestion attracted the attention of Avery Jones and Hirschowitz and the latter developed a simple prototype fibreoptic gastroscope, which was first demonstrated in 1957. However, the necessary backing for the further development of this instrument was not forthcoming in Britain, and the idea was taken up first in America and later in Japan.

It is interesting to look at the history of bronchoscopy in Japan, since that country now produces most of the fibreoptic bronchoscopes used throughout the world. The first epoch began in 1907, when Professor I. Kubo at Kyushu University studied under Gustav Killian (1860-1921) and then succeeded in removing a foreign body from a four-year-old boy. The second epoch was when in 1934 Professor J. Ono of Keio University went to the USA and studied under the guidance of Chevalier Jackson. On returning to Japan, he tried to introduce broncho-oesophagology by organizing various seminars; later the Japanese Broncho-Oesophagological Society was established in 1949.

The third epoch began in 1966 (forty-first year of Showa) when Ikeda's flexible bronchofibrescope was developed respectively by Machida and Olympus. Ikeda introduced flexible bronchofibrescopy for the first time ever at the World Congress on Diseases of the Chest held in Copenhagen in August of 1966.

Subsequently the fibrescopes have been developed progressively with better lenses and mobility and wider internal channels. However, the glass fibres themselves have virtually reached the limit of development. Below 5µm diameter the glass fibre begins to lose light of various wavelengths, and above 5µm the image becomes grainy and unclear, while the fibrescope becomes too wide for convenience. Further development may witness the incorporation of electronics into the instruments.

Initially the Japanese inserted the fibreoptic bronchoscope through a flexible endotracheal tube or rigid bronchoscope, but this was quickly superseded by the transnasal or transoral approach advocated by Smiddy and colleagues in 1971. Likewise, general anaesthesia was overtaken by local anaesthesia. At a stroke, this made fibreoptic bronchoscopy a simple technique to be carried out in a side-room or at the bedside, in contrast to the general anaesthesia and theatre requirements of rigid bronchoscopy. This effectively opened the door to physicians, and enabled them to bronchoscope patients readily for a wide variety of non-surgical disorders.

Subsequently, fibreoptic bronchoscopy has virtually replaced rigid bronchoscopy. The number of fibreoptic bronchoscopies has risen sharply and in many centres has surpassed the number of rigid bronchoscopies by several fold. Although a rigid bronchoscope is often held in readiness, many fibreoptic bronchoscopists will never have need to use one, and may find it difficult to keep in practice.

This book describes in detail the various fibreoptic bronchoscopic techniques, such as transbronchial biopsy and bronchoalveolar lavage, and examines the situations in which the fibreoptic bronchoscope is used, whether for diagnosis or therapy. Case studies throughout the text give real examples of presentations that have been followed by fibreoptic bronchoscopy.

1. Techniques

S. W. Clarke MD FRCP FCCP

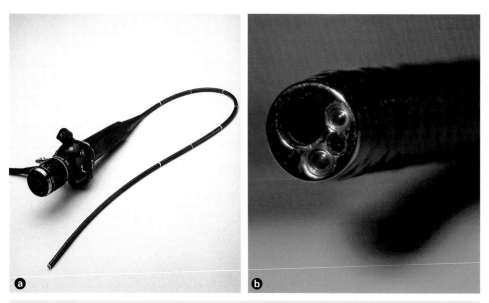

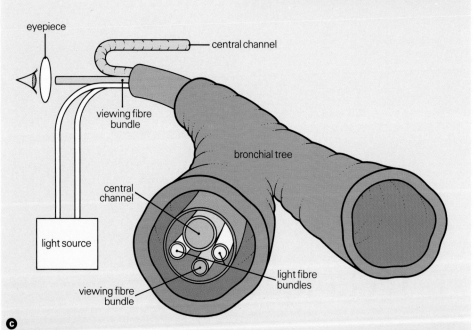

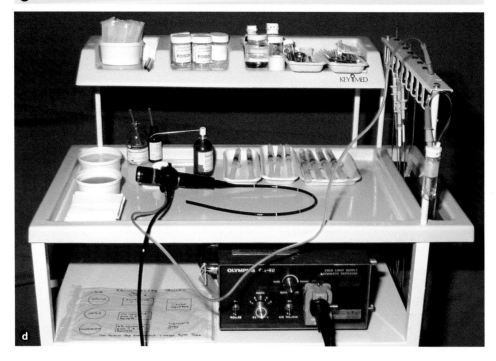

Fig. 1.1 The fibreoptic bronchoscope. Photograph (a) shows the whole instrument from butt to tip and (b) shows the fibrescope tip. The diagram (c) illustrates the various parts and view (d) shows the bronchoscopy trolley with fibrescope, sputum trap on the right and light source on the bottom shelf.

1.2

GENERAL METHODS

Position of the Bronchoscopist and Patient

The fibreoptic bronchoscope (Fig. 1.1) can be used in many ways, depending on individual and sometimes patient preference (Fig. 1.2).

The most important concept here is that of ignoring the position of the patient and operator while focusing on the anatomy of the airways, which is unchangeable, compared with the external orientation, which is not. Thus the experienced bronchoscopist should be able to bronchoscope a patient in any position. This is quite the reverse of rigid bronchoscopy, where the patient and bronchoscopist are always in the same relative position; a point which is related to the technique of passing a straight, rigid tube around the right-angled bend of the oropharynx.

It should also be pointed out that the view seen through the fibreoptic bronchoscope usually has a different orientation to that seen through the rigid bronchoscope, because of the different positioning of

patient and bronchoscopist. This is reflected in many of the photographs shown in this book which have mainly been taken through the fibreoptic bronchoscope.

The techniques of individual bronchoscopists vary widely. This is not important provided that bronchoscopy is completed satisfactorily, with a good view of the upper airways and tracheobronchial tree and adequate specimens obtained. Most bronchoscopists favour the direct transnasal approach with the patient supine at forty-five degrees (Fig. 1.3), while the bronchoscopist stands or sits to the patient's right, facing the patient. This enables the bronchoscopist to observe the patient closely and is very satisfactory.

The transoral approach (Fig. 1.4), which requires a mouthguard to prevent biting, is used if the nasal passage is too small for comfortable access. This occurs in about five per cent of adults. The only disadvantage is that the fibrescope tip tends to slide around the pharynx and causes gagging and retching, more so than with the

Fibreoptic Bronchoscopic Techniques

	routine bronchoscopy including bronchoalveolar lavage (BAL)		transbronchial biopsy
route	nasal or oral	oral	nasal or oral
anaesthesia	local	general	local or general
patient's position	supine supine at 45° sitting (e.g. in dental chair)	supine	supine
bronchoscopist's position	head side (usually patient's right)	side (anaesthetist at head)	head side

Fig. 1.2 Fibreoptic bronchoscopic techniques.

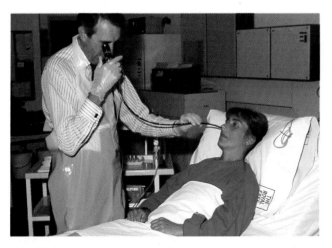

Fig. 1.3 Bronchoscopic technique. The direct transnasal approach with the patient supine at forty-five degrees and the bronchoscopist on the right-hand side.

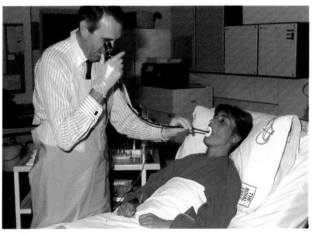

Fig. 1.4 Bronchoscopic technique. The transoral approach with the patient supine at forty-five degrees and the bronchoscopist on the right-hand side.

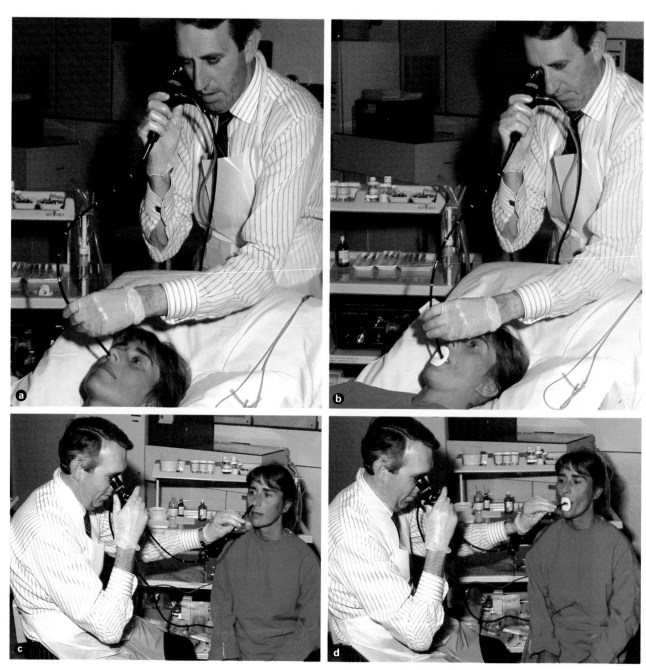

Fig. 1.5 Bronchoscopic technique. The following views are shown: the transnasal and transoral approaches with the patient supine and the bronchoscopist behind the head, views (a) and (b) respectively; the transnasal and transoral approaches with the patient sitting and the bronchoscopist facing on the right-hand side, views (c) and (d) respectively.

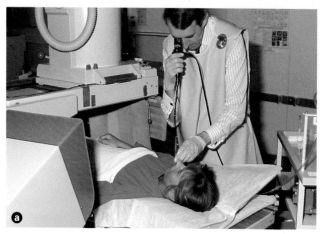

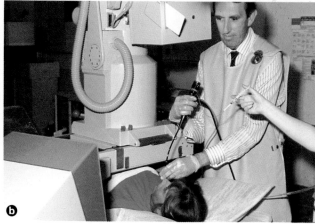

Fig. 1.6 Transbronchial biopsy procedure. The transnasal approach with the patient supine is employed (a); X-ray equipment and monitor are visible. View (b) shows the positioning of the biopsy forceps, held by an assistant, while screening.

transnasal route. This abates after the fibrescope has passed the larynx. Some bronchoscopists have stated the need for an endotracheal tube to 'control' the airway. This is unpleasant for the patient and largely unnecessary, but does allow multiple insertions for taking samples. Further approaches are shown in Figure 1.5.

If a general anaesthetic is required, usually because of intense anxiety on the part of the patient, a suitable endotracheal tube is inserted. A T-piece at the proximal end permits the fibrescope to be passed while ventilation proceeds via the other limb. Special plastic T-pieces are available, but a metal one with a snug fit usually suffices with little anaesthetic blow-back. With this technique it is difficult to visualize the larynx and upper trachea well, and furthermore the full mobility of the fibrescope is curtailed by the endotracheal tube. There is the advantage, however, that the fibrescope can be withdrawn and reinserted at will. The post-anaesthetic care must be of greater length and more rigorous than with local anaesthetic. Transbronchial biopsy is usually performed under local anaesthetic with the patient supine to permit screening. The forceps can then be placed more accurately and the likelihood of pneumothorax by pleural biopsy is reduced (Fig. 1.6).

Handling the Fibrescope

The butt of the instrument may be held by the dominant hand, with the thumb on the toggle and the index finger on the sucker (Fig. 1.7). The other hand is then used to control the shaft entering the patient's nose. With practice it can be achieved with the hands the reverse way around, and again the bronchoscopist can exercise his own preference. It is best to have an assistant to help pass and manipulate the biopsy forceps and brush.

Gentle handling of the fibrescope is of vital importance; acute flexing, trapping (in the case), banging or dropping are to be avoided, since all may damage the image bundle

and impair the view with a myriad of black spots from broken fibres.

Patient Assessment

The patient is carefully assessed beforehand, both clinically and with chest radiographs. The relevant radiographs are put on a viewing box at the time of bronchoscopy. The age, diagnosis, clinical severity of disease and the presence of airway obstruction are noted, together with spirometry or full pulmonary function tests; if serious hypoxia or hypercapnia is suspected, arterial blood gases are checked. If the patient either is becoming or is at risk of becoming hypoxic, oxygen is supplied at a rate of at least two litres per minute through a contralateral nasal prong. The incidence of hypoxia during fibreoptic bronchoscopy is usually small (Albertini et al. 1974).

Anaesthesia and Fibrescope Insertion

The patient should fast for four hours prior to anaesthesia. Pre-medication usually consists of papaveretum (10-20mg) and scopolamine (0.2-0.4mg) intramuscularly, one hour previously. Papaveretum sedates, but importantly also suppresses coughing; scopolamine blocks vagal activity and reduces secretions. In aged patients or in patients with severe airways obstruction, both may need omitting and diazepam (5-10mg intramuscularly or intravenously) may be given as a substitute. This is less satisfactory by itself. Other variations have been suggested including morphine and benzodiazepine (Simpson et al. 1986).

A simple but clear explanation of the procedure is given to the patient beforehand to allay anxiety and help co-operation, since the patient is conscious throughout although often drowsy. The procedure may be carried out either in a bronchoscopy suite, clinic, side room or at the bedside. Many are now done as day cases (outpatients).

Fig. 1.7 Handling the fibrescope. The butt of the instrument is usually held by the dominant hand, with thumb on the toggle and index finger on the sucker as shown.

The patient should lie or sit comfortably. The widest nostril is anaesthetized with lignocaine spray; approximately four puffs with the patient sniffing it back (Fig. 1.8). This is 'sharp' to the nose and has an acrid, unpleasant taste but is highly effective within seconds, often causing some coughing, indicating deposition down to the larynx. The pharynx is then anaesthetized with two or three puffs of lignocaine which can be swallowed (Fig. 1.9). The aerosol applicator is then placed over the back of the protruding tongue and about

four puffs are sprayed in the direction of the larynx with the patient breathing deeply. Again this may cause some coughing.

The fibrescope is checked to ensure that it is in good working order and that the locking lever is placed to off; the lenses are then cleaned and polished to give a good view and the focus is adjusted. The shaft is liberally coated with lignocaine gel; gel may be placed in the nostril.

The fibrescope is passed under direct vision through the anterior nares (Fig. 1.10) and then via the middle

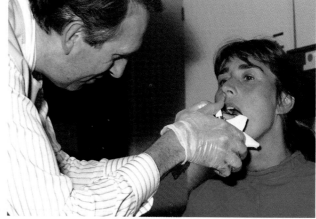

Fig. 1.8 Anaesthetizing the widest nostril with lignocaine spray prior to bronchoscopy.

Fig. 1.9 Anaesthetizing the pharynx and larynx with lignocaine spray prior to bronchoscopy.

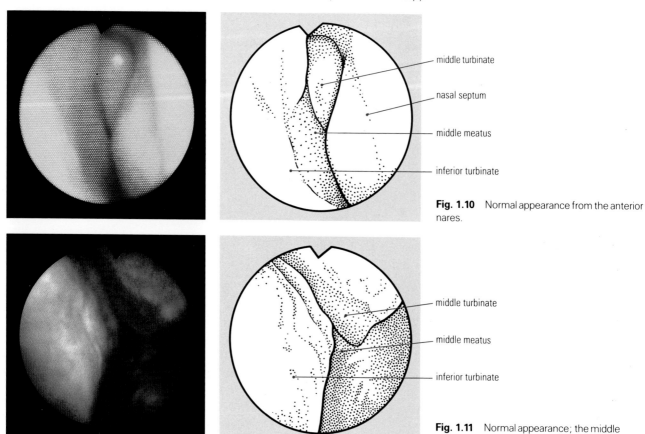

Fig. 1.10 Normal appearance from the anterior nares.

middle turbinate

nasal septum

middle meatus

inferior turbinate

Fig. 1.11 Normal appearance; the middle meatus.

middle turbinate

middle meatus

inferior turbinate

meatus (Figs 1.11 and 1.12) and occasionally, the inferior meatus. This is done very gently with a minimum of pushing, as mucosa usually yields easily while bony structures do not and are painful. Having traversed the nose, the tip is gently angled about ninety degrees down through the nasopharynx (Figs 1.13 to 1.15), where-

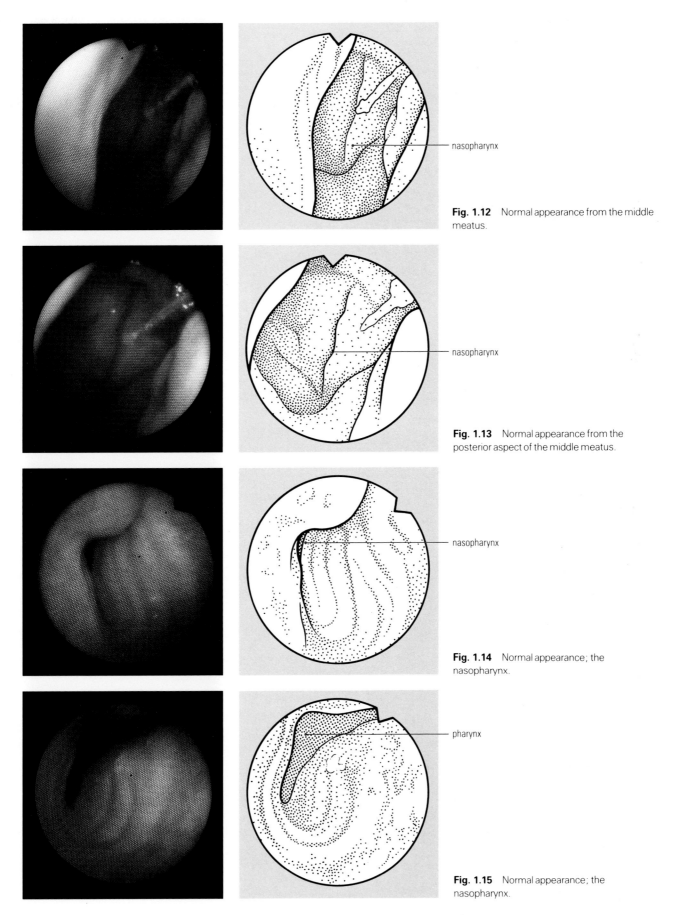

Fig. 1.12 Normal appearance from the middle meatus.

Fig. 1.13 Normal appearance from the posterior aspect of the middle meatus.

Fig. 1.14 Normal appearance; the nasopharynx.

Fig. 1.15 Normal appearance; the nasopharynx.

upon the epiglottis becomes visible (Fig. 1.16). The tip is then slipped behind the epiglottis where it falls naturally onto the larynx (Figs 1.17 to 1.19). At this stage, gagging may deposit the tip in the pyriform fossa, a most efficient defence, but this is easily overcome by withdrawal, identifying the larynx and replacing the fibrescope.

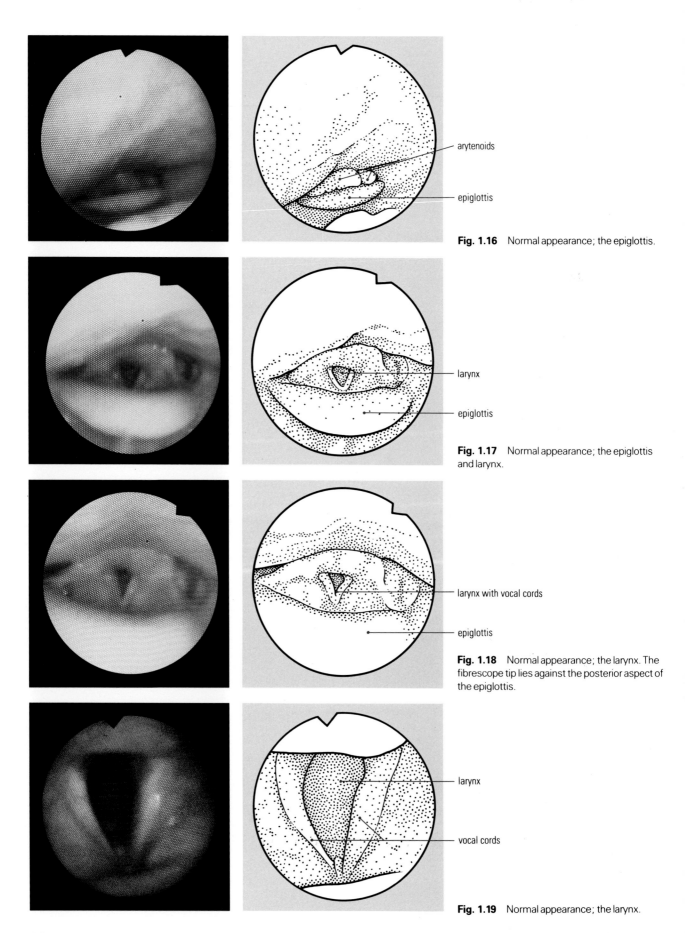

Fig. 1.16 Normal appearance; the epiglottis.

Fig. 1.17 Normal appearance; the epiglottis and larynx.

Fig. 1.18 Normal appearance; the larynx. The fibrescope tip lies against the posterior aspect of the epiglottis.

Fig. 1.19 Normal appearance; the larynx.

With the tip held steady, approximately 1cm above the vocal cords, 2ml of four per cent lignocaine is sprayed directly onto the cords through the instrument's centre channel. This often causes a brief bout of coughing, about which the patient is forewarned. A further 2ml of four per cent lignocaine is given, and the fibrescope held still for a minute at least to await full anaesthesia. Then, with the patient concentrating on deep breathing, the fibrescope is gently advanced in the mid-line to pass through the posterior and widest aspect of the vocal cords with a 'no touch' technique. Often this can be accomplished without any coughing; on occasion, there is transient coughing, insufficient to trouble either bronchoscopist or patient. The trachea is anaesthetized with an injection of 2ml of two per cent lignocaine, and subsequent injections are made to both bronchial trees, as required, to suppress coughing.

On average approximately four to six 2ml aliquots of two per cent lignocaine are required for satisfactory cough suppression; this lasts between twenty and thirty minutes in all. It may be replenished up to a total dose of 400mg but no more, otherwise in theory toxic blood levels might result. In practice, lignocaine toxicity rarely, if ever, occurs (Efthimiou et al. 1982).

Inspection Procedure

As the fibrescope is inserted, the vocal cords are carefully inspected for normality and mobility; an immobile cord indicates recurrent laryngeal palsy, usually on the left side. The sub-laryngeal area can next be inspected immediately below the cords, and this area is much better seen with the fibrescope than with the laryngoscope or rigid bronchoscope. The trachea is then inspected, particular note being made of the surface mucosa, which is smooth and pink with tracheal rings, indentations and deviation. The carina should be sharp, clearcut and mobile, mobility or fixity being observed during deep breathing or coughing, when the trachea and main bronchi close down and the carina contracts.

The right bronchial tree is usually examined first, the fibrescope being slid down to the right middle lobe medial and lateral segments, then to the lower lobe anterior, medial, lateral and posterior basal segments, then the apical which lies posteriorly, opposite the right middle lobe, and finally the upper lobe orifice, opposite and just below the carina, with the anterior, posterior and apical segmental orifices. On the left side, the lingular and upper lobe orifices can be inspected first, and then the lower lobe orifices in the same sequence.

On occasion, it is advisable to go straight to the site of the anticipated lesion to identify it and take a biopsy. Nevertheless, a full general inspection should always be done. The plan of inspection may vary depending on the bronchoscopist's preference. Once inserted, the fibrescope is not withdrawn until the examination has been completed satisfactorily. With experience it is easy to keep the tip clean with suction and saline, by wiping on the mucosa, by asking the patient to cough forcibly and by avoiding leaving the tip where it can become fogged with blood, particularly after biopsy. It is the mark of inexperience if this cannot be accomplished.

OBTAINING SAMPLES
Sputum Trap

A sputum trap can be placed in the suction line (Fig. 1.20), after the instrument has been inserted, to sample uncontaminated secretions or lavage fluids. Several traps may be used when aspirating different sites, for example, when seeking occult bronchial carcinoma by cytology. The trap specimen is routinely sent for both smear and culture, including tests for tubercle and fungi, as appropriate, and also for cytology.

Forceps Biopsy

A visible lesion or abnormal mucosa may be biopsied by one of the several types of forceps available (Fig. 1.21). Normal bronchial mucosa is fairly tough and the forceps tend to slip on the submucosa, merely tearing the surface

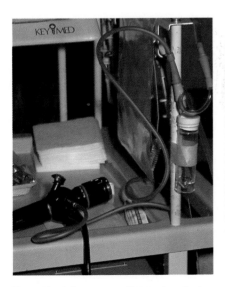
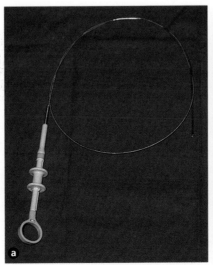

Fig. 1.20 A Sputum trap. This is placed in the suction line after the instrument has been inserted and collects uncontaminated secretions and lavage fluid.

Fig. 1.21 One of the several types of biopsy forceps available for the fibreoptic bronchoscope. The whole instrument (a) and a close-up of the forceps (b) are shown.

mucosa; thus deeper parts of the wall are not included. In the case of a tumour, it is best to take between three and five biopsies from around the edge, and not from the centre which may be necrotic, in order to make a firm histological diagnosis including cell type.

The forceps should be passed with the fibrescope slightly withdrawn and the tip straightened; otherwise there may be difficulty and they may damage the distal fibres and internal channel. The jaws of the forceps are pushed out sufficiently to clear the tip of the fibrescope, which is then moved into position. The jaws are then opened and the biopsy is taken under clear vision and with pinpoint accuracy. With the blades closed a little tug will avulse the specimen. The forceps are then withdrawn and the sample is shaken or teased off into ten per cent formalin. The forceps are finally washed in sterile saline and are then ready for further use. Once a biopsy has been taken, the fibrescope tip should be withdrawn a centimetre or so to avoid a blood smear or clot on the tip. An assistant should handle the forceps and biopsy.

Brush Biopsy

A shielded brush (Fig. 1.22) is preferred to brush either visible lesions, segmental lesions or lesions identified under the image intensifier (Wimberley et al. 1979). Upon withdrawal, the brush is smeared on a microscope slide which is either fixed immediately or agitated in sterile normal saline or culture medium. Repeat brushings can be made. Brush biopsy can be used alone or as an adjunct to forceps biopsy, whereby a wider area of bronchial mucosa can be sampled with minimal trauma.

Needle Biopsy

A needle may be used to do an aspiration biopsy through the fibrescope of, for example, hilar or paratracheal glands. This technique is of growing importance, giving a high yield (Wang & Terry 1983). By comparison, neither forceps nor brush biopsy penetrate the bronchial wall (see also *Chapter 3*).

Transbronchial Biopsy

This is now performed routinely with forceps passed through the fibreoptic bronchoscope (Fig. 1.23 – see also *Chapter 4*). It is a simple, safe method of sampling peripheral lung tissue either in diffuse or localized lung disease. The procedure is best performed under the fluoroscope and image intensifier so that pinpoint accuracy may be achieved, though in diffuse lung disease this is not mandatory. It now takes its place with percutaneous lung biopsy, which may be favoured for localized lesions nearer to the lung periphery than the hilum, high-speed trephine biopsy, which has now gone out of vogue, and open lung biopsy, which yields a much larger sample of representative tissue and must remain the gold standard by which others are judged.

Preparation

Spirometry and arterial blood gases are checked as necessary before the operation. Platelets and bleeding and clotting times are routinely checked and two units of blood cross-matched in case of haemorrhage; it is rarely needed. Pre-medication and local anaesthesia are identical to that used for routine bronchoscopy.

Technique

With diffuse lung disease it is often convenient to biopsy the right lower lobe, though most parts of the lungs can be reached. The fibrescope is placed at the orifice of an appropriate segmental bronchus visually, thereafter the forceps are observed on the screen with the image intensifier. They are advanced until resistance is felt upon impaction in a peripheral bronchus of between 2-3mm in diameter. Pain felt either in the chest, shoulder tip or upper abdomen indicates pleural irritation and is an indication for resiting the forceps; a pneumothorax may otherwise result. Once the forceps are in position, they are withdrawn 1–2cm and opened. With the patient taking a deep inspiration, they are advanced as far as gentle pressure allows and the jaws closed at the end of the next expiration.

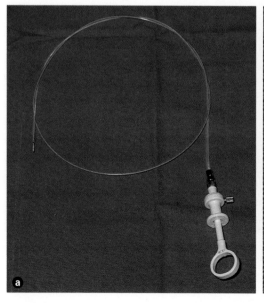

Fig. 1.22 Shielded biopsy brush for use with the fibreoptic bronchoscope. The whole instrument (a) and a close-up of the brush (b) are shown.

If a distal subcarina and the adjacent lung tissues have been grasped by the forceps, an initial light resistance to traction will be felt and lung movement may be seen on the fluoroscope. Continuing traction will avulse the specimen. Fibrotic lung tissue may be particularly tough.

The specimen is either shaken into ten per cent formalin solution for histology, saline for microbiology or transport medium for virology. If a specimen contains lung tissue it appears 'fluffy' and floats. Further biopsies of between five and ten in number may be obtained from adjacent sites in the same lung. Any bleeding, which is usually trivial, may be cleared by lavage and aspiration.

An identical technique is used for biopsy of localized lesions, though the accurate siting of the forceps is more important and will require lateral as well as anteroposterior screening. At the end of the procedure, check screening for pneumothorax is routine. In the event of substantial haemorrhage, the fibrescope should be held *in situ*, the blood aspirated, the patient turned with the affected side down and head down and oxygen given. Rarely, balloon tamponade via the centre channel and blood transfusion may be required. Side-effects are in reality relatively uncommon (Credle et al. 1974).

Indications

Transbronchial biopsy may be useful in any patient with diffuse lung disease. It usually gives a clear answer in sarcoidosis and may do so in alveolar proteinosis (when lavage also helps), eosinophilic granuloma and lymphomatous or carcinomatous infiltration. It is particularly useful in diagnosing *Pneumocystis carinii* pneumonia (PCP), which is often seen in both immunocompromised patients after transplants and patients with the acquired immune deficiency syndrome (AIDS); lavage is also useful here. It is less satisfactory with fibrosing alveolitis where the tissue volume may be too small to be representative and may show crush artefact. As mentioned previously, it may be useful for discrete lesions, such as tumours, granulomata and hamartomas.

Bronchoalveolar Lavage

This technique for washing cells from the periphery of the lung is usually performed for assessment of diffuse parenchymal lung disease during routine bronchoscopy (Turner-Warwick & Haslam 1986). The type of lung disease includes sarcoidosis, cryptogenic fibrosing alveolitis (CFA) and others such as occupational dust diseases, haemosiderosis, alveolar lipoproteinosis and eosinophilic granuloma. The patient is fully assessed prior to bronchoscopy, with a current chest radiograph and lung function tests, including arterial blood gases. There are usually several contraindications, including the presence of relevant heart disease, a forced expiratory volume (FEV_1) of less than one litre, arterial PaO_2 less than 9.3kPa (70mmHg), recent infection and age over sixty-five years, though many would consider the latter too stringent.

The fibrescope is gently wedged into a basal bronchus, often in the lateral segmental orifice of the right lower lobe, though some bronchoscopists use the middle lobe or lingula. 60ml aliquots of buffered saline are introduced, pre-warmed to 37°C. The fluid is gently aspirated to minimize trauma into a siliconized glass bottle on ice. This is repeated three times, up to 180ml, and if 100ml or

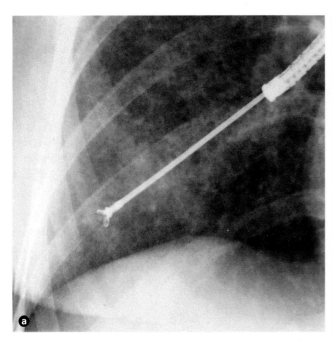 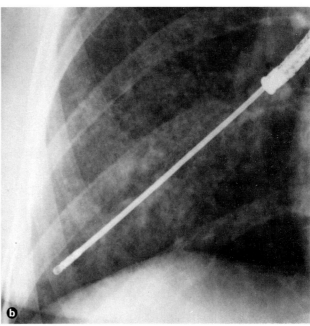

Fig. 1.23 Radiographs of transbronchial biopsy procedure in the right lower lobe. The biopsy forceps can be seen protruding beyond the end of the fibrescope with jaws open in view (a) and closed in (b).

more is recovered, the procedure is terminated. Otherwise further 60ml aliquots are instilled up to a maximum of 240-300ml.

The sample is taken to the laboratory immediately. The patient receives supplemental oxygen at a rate of 4 litres/min during the procedure and this is continued for four hours thereafter, to avoid hypoxia, since lavage induces a mean fall of 3kPa (22.5mmHg).

Complications of this procedure are usually mild; they are post-lavage pyrexia ($\geq 1°C$ rise), post-lavage fall in peak expiratory flow rate (PEFR, $\geq 20\%$ fall) and the occasional bradycardia and haemoptysis. Of these, pyrexia alone is related to lavage and may be seen in up to twenty-six per cent of patients whereas the rest are seen also with simple bronchoscopy. Further, pyrexia is related to the volume of lavage fluid instilled; with 240ml or less, the incidence of pyrexia is ten per cent and 240ml or more forty per cent. Prior treatment with prednisolone and immunosuppressive drugs is associated with an increased incidence.

Bearing all these points in mind, bronchoalveolar lavage is now an established procedure which can be used for the appropriate clinical conditions and with suitable care (Crystal et al. 1986). Further details are given in *Chapter 4*.

PAEDIATRIC BRONCHOSCOPY

Until recently, the rigid bronchoscope has been the instrument of choice in children. However, there are now new techniques which allow even small children of a few years of age to be safely bronchoscoped using the paediatric fibreoptic instrument (Warner et al. 1981). One technique involves ventilation of the child by an oxygen venturi jet through a small nasal catheter. This lies with its tip in the mid-tracheal region and is performed under general intravenous anaesthetic. The paediatric bronchoscope is then passed alongside the catheter allowing a conventional examination.

BRONCHOSCOPY IN THE INTENSIVE CARE UNIT

Bronchoscopy in the intensive care unit is relatively simple, particularly if the patient has an endotracheal or tracheostomy tube *in situ*. The fibrescope either can be passed through the open limb of a T-tube, while ventilation continues with an appropriate oxygen/air mixture, allowing for some leakage around the instrument's shaft or it can be passed by using an appropriate endotracheal tube adaptor. Lignocaine is rarely required since the cough reflex is usually suppressed. Provided the period of insertion is brief, lasting only a few minutes, and the patient is observed for tachycardia and cyanosis, there are no serious side-effects. Multiple insertions for sputum aspiration and the full range of bronchoscopic techniques may be safely undertaken.

REFERENCES

Albertini R, Harrel JH, Kurihara N, Moser KM (1974). Arterial hypoxaemia induced by fiberoptic bronchoscopy. *Journal of the American Medical Association,* **230,** 1667-1668.

Credle WF, Smiddy JF, Elliot RC (1974). Complications of fiberoptic bronchoscopy. *American Review of Respiratory Disease,* **109,** 67-72.

Crystal RG, Reynolds HY, Kalica AR (1986). Bronchoalveolar lavage. *Chest,* **90,** 122-131.

Efthimiou J, Higenbottam T, Hold D, Cochrane GM (1982). Plasma concentrations of lignocaine during fibreoptic bronchoscopy. *Thorax,* **37,** 68-71.

Ikeda S (1974). Atlas of Flexible Bronchofiberscopy. Igaku Shoin Ltd, Tokyo: 230pp.

Oho K, Amemiya R (1984). Practical Fiberoptic Bronchoscopy. 2nd edition. Igaku Shoin Ltd, Tokyo: 218pp.

Simpson FG, Arnold AG, Purvis A, Belfield PW, Muers MF, Cooke NJ (1986). Postal survey of bronchoscopic practice by physicians in the United Kingdom. *Thorax,* **41,** 311-317.

Turner-Warwick ME, Haslam PL (1986). Clinical applications of bronchoalveolar lavage: an interim view. *British Journal of Diseases of the Chest,* **80,** 105-121.

Wang KP, Terry PB (1983). Transbronchial needle aspiration in the diagnosis and staging of bronchogenic carcinoma. *American Review of Respiratory Disease,* **127,** 344-347.

Warner JO, Clarke SW, Scallan MJH (1981). Fibre-optic bronchoscopy in small children. In: Nakhosteen JA, Maassen W eds, Bronchoscopy: Research, Diagnosis and Therapeutic Aspects. Martinus Nijhoff, The Hague: 511-513.

Wimberley N, Faling LJ, Bartlett JG (1979). A fiber-optic bronchoscopic technique to obtain uncontaminated lower airway secretions for bacterial culture. *American Review of Respiratory Disease,* **119,** 337-343.

Zavala DC (1978). Flexible Fiberoptic Bronchoscopy. University of Iowa, Iowa: 165pp.

2. Haemoptysis with a Normal Chest Radiograph

S. W. Clarke MD FRCP FCCP

INTRODUCTION

There are a number of situations when a patient may present with respiratory symptoms for which broncho-scopy is performed, despite a normal chest radiograph. The most common of these symptoms is haem-optysis and indeed there are few other instances where symptoms such as breathlessness, chest pain or persistent sputum, associated with a normal chest radiograph, require bronchoscopic investigations. This chapter, therefore, concentrates on haemoptysis and illustrates a number of examples in which this symptom occurs in the absence of radiological abnormality.

HAEMOPTYSIS

Haemoptysis is a common symptom in chest disease and accounts for between ten and thirty per cent of bronchoscopies in most centres. The first important step in diagnosing the cause is to ascertain whether or not the blood is expectorated from the tracheobronchial tree. In some instances, the blood may be aspirated from a nosebleed, of which there may be a history, or arise from bleeding elsewhere, such as the teeth or stomach (haematemesis). A careful history and physical examination usually clarifies the situation; the chest radiograph may, however, be clear.

The majority of the causes of haemoptysis are fairly trivial. Large analyses have found that inflammatory conditions such as acute or chronic bronchitis are the commonest causes, whereas more serious causes such as cancer and tuberculosis are relatively uncommon

(Fig. 2.1). Rarer causes include Goodpasture's syndrome, haemosiderosis, telangiectasia, bleeding disorders, aspergilloma and cardiac failure with pulmonary congestion. All these causes are self-evident on clinical appraisal when supported by chest radiograph changes and other positive investigations.

In a number of cases, estimated at between twenty and thirty per cent, however, the cause of haemoptysis is unclear and this is particularly important where symptoms are recurrent. Undiagnosed carcinoma is then a strong possibility in both males and females older than forty years of age with a smoking history of more than forty pack years and haemoptysis of a week or more. Other causes include bronchiectasis where the changes are insufficient to show on plain chest radiography, although some would dispute the absence of signs given close, expert scrutiny of the radiographs. Pulmonary embolism is a frequent cause of haemoptysis and on occasion neither the clinical features nor the chest radiograph will give a clear answer. However, lung ventilation-perfusion scans almost always confirm the diagnosis thereby obviating the need for bronchoscopy. Finally, early endobronchial tuberculosis may occasion-ally present with haemoptysis, prior to the appearance of chest radiograph changes, and the occasional inhaled radiolucent foreign body or bronchial adenoma may also present with haemoptysis and a clear chest radiograph.

Where cancer is suspected, but not visible on a chest radiograph, three sputum samples are usually tested

Investigation of Causes of Haemoptysis

upper respiratory infection	24%
bronchitis	17%
bronchiectasis	13%
pulmonary tuberculosis	
active	5%
quiescent	6%
pneumonia	6%
cardiovascular disease	
hypertension	2%
mitral valve disease	2%
carcinoma	2%
others	2%
no abnormality found	21%

Fig. 2.1 The causes of haemoptysis. Diagnosis of 324 patients presenting with haemoptysis (Johnston et al. 1960).

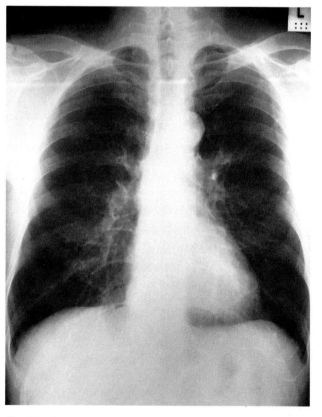

Fig. 2.2 Case study 1. Normal chest radiograph despite recurrent haemoptysis.

cytologically and it is then that bronchoscopy is performed. The most likely site of the tumour is in the central airways, as it would otherwise probably show on the chest radiograph. Occult lung cancer usually refers to the situation where chest radiography is clear but cytology is positive. Here considerable care with selective brushing and lavage is required for localization of the cancer. In general, the combination of a clear chest radiograph and a negative bronchoscopy virtually rules out lung cancer and this provides welcome reassurance for the patient with haemoptysis.

Case Study 1

A sixty-one-year-old-man, who was a heavy smoker of greater than forty pack years, presented with recurrent haemoptysis over a period of four weeks. Clinical examination and a chest radiograph (Fig. 2.2) were normal. Fibreoptic bronchoscopy (Fig. 2.3) showed tumour infiltrating the carina, right main and upper lobe bronchi; histological examination

of the biopsy showed a squamous cell carcinoma which responded well to radiotherapy.

Case Study 2

A seventy-year-old man, who was an ex-smoker of thirty pack years, presented with haemoptysis which he had experienced for one week; his chest radiograph was clear. Fibreoptic bronchoscopy showed a small red lesion lying at the origin of the apicoposterior segmental bronchus of the right lower lobe; it had the appearance of a small haemangioma (Fig. 2.4). Trap specimen was negative and brush biopsy showed no sign of malignancy.

Case Study 3

A thirty-two-year-old man, who was an ex-smoker, presented with haemoptysis and a clear chest radiograph. Fibreoptic bronchoscopy showed a 'cherry-like' tumour in the left lower lobe which proved to be a bronchial adenoma (Fig. 2.5). He was cured by resection.

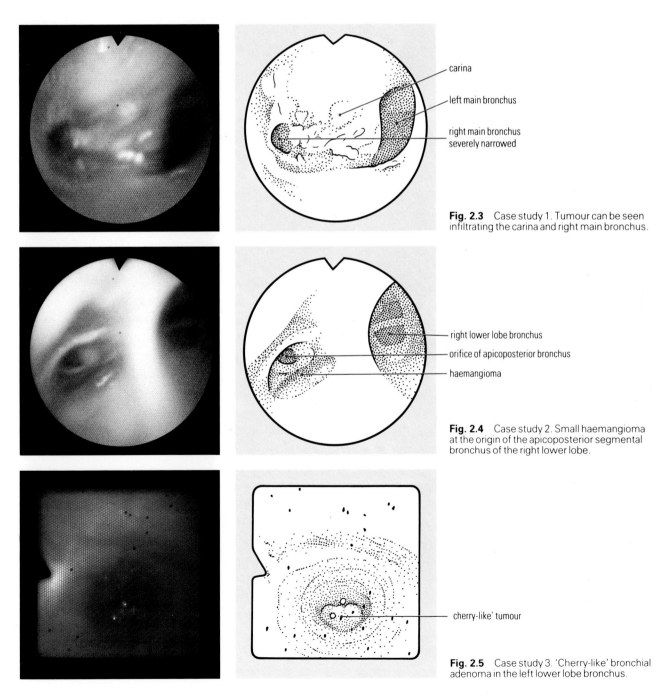

carina

left main bronchus

right main bronchus severely narrowed

Fig. 2.3 Case study 1. Tumour can be seen infiltrating the carina and right main bronchus.

right lower lobe bronchus

orifice of apicoposterior bronchus

haemangioma

Fig. 2.4 Case study 2. Small haemangioma at the origin of the apicoposterior segmental bronchus of the right lower lobe.

'cherry-like' tumour

Fig. 2.5 Case study 3. 'Cherry-like' bronchial adenoma in the left lower lobe bronchus.

Where bronchiectasis is suspected, both chest radiography and bronchoscopy may be negative while bronchography, often performed through the fibrescope, is positive in fifteen per cent of such cases.

Case Study 4

A forty-nine-year-old man with haemoptysis had a clear chest radiograph and bronchoscopy was negative. On selective bronchography (Fig. 2.6), bronchiectasis was revealed in the right lower zone.

Overall, only a small number of patients with haemoptysis and a normal chest radiograph need to be bronchoscoped to rule out carcinoma, provided the following criteria are taken into consideration: (i) age greater than forty; (ii) smoking history greater than forty pack years; (iii) haemoptysis of more than one week duration. Nevertheless, if doubt about carcinoma or the risk of some other disease remains either in the doctor's or patient's mind, then bronchoscopy should be performed, particularly since the reassurance of a negative finding is considerable and the risks of fibreoptic bronchoscopy under local anaesthesia are negligible.

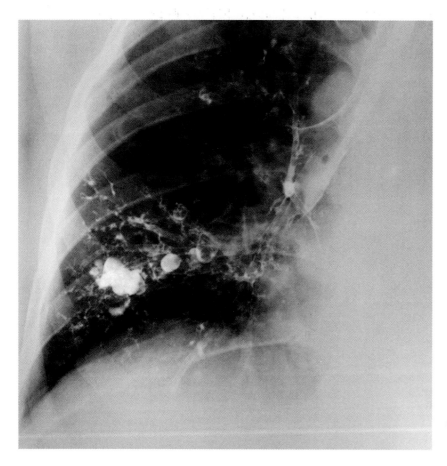

Fig. 2.6 Case study 4. Bronchogram of the right lung showing bronchiectasis in the right lower zone.

REFERENCES

Jackson CV, Savage PJ, Quinn DL (1985). Role of fiberoptic bronchoscopy in patients with haemoptysis and a normal chest roentgenogram. *Chest,* **87,** 142-144.

Johnston RN, Lockhart W, Ritchie RT, Smith DH (1960). Haemoptysis. *British Medical Journal,* **1,** 592-595.

Jones DK, Cavanagh P, Shneerson JM, Flower CDR (1985). Does bronchography have a role in the assessment of patients with haemoptysis? *Thorax,* **40,** 668-670.

3. Investigation of
the Mass Lesion
on Chest Radiography

D. E. Stableforth MA MB BCh FRCP

INTRODUCTION

The advent of the fibreoptic bronchoscope has revolutionized the clinical approach to patients with one or several masses detected by chest radiography. It has enabled both physicians and surgeons to gain access to the bronchial tree for the application of a number of diagnostic techniques. There is a wide range of possible causes for intrathoracic masses, be they visible in the mediastinum, at the hilar regions, in the airways, the lung fields, the pleural space or even on the chest wall (Fig. 3.1).

Once a mass has been identified using radiography, there is no substitute for a meticulously taken history, noting symptoms, past medical events, smoking, possible industrial dust exposure, exercise tolerance and general health. This should then be followed by a full clinical examination: initial routine screening investigations such as sputum culture, which includes a search for tubercle bacilli, sputum cytology, electrolyte analysis, liver function tests, erythrocyte sedimentation rate (ESR) and *Aspergillus* precipitins, may all give important clues to the cause of an abnormality on the chest radiograph.

For logical management, either a histological diagnosis or an accurate prediction of it based on clinical investigations is required. This has been facilitated since the early 1970's by the fibreoptic bronchoscope which enables easy and safe access to the bronchial tree. It allows endoscopic examination, aspiration of secretions and lavage fluid for bacterial culture and cytology, bronchial and transbronchial biopsies, transbronchial needle aspiration and selective brushing of visible lesions or those beyond the range of vision through the fibrescope.

DIAGNOSTIC METHODS

There are a number of diagnostic methods available for the investigation and assessment of mass lesions (Fig. 3.2).

Radiology

A routine posteroanterior (PA) chest radiograph is not usually sufficient to identify precisely the anatomical site of a mass and lateral views and conventional or computerized axial tomograms will usually be necessary. Although

Lesions Causing a Mass on Chest Radiography

neoplastic	infective	miscellaneous
malignant	**bacterial**	sarcoidosis
primary bronchial carcinoma	pneumonia	rheumatoid nodules
bronchial adenoma	lung abscess	pseudolymphoma
lymphoma	empyema	Wegener's
plasmacytoma		granulomatosis
thymoma	**tuberculous**	bronchocentric
germinoma	tuberculoma	granulomatosis
metastatic carcinoma		
	fungal	
benign	aspergilloma	
neurofibroma	allergic aspergillosis	
hamartoma	histoplasmoma	
thymoma	mycetoma	

Fig. 3.1 Lesions which may cause a mass on the chest radiograph.

Diagnostic Methods for Assessment of Mass Lesions

fibreoptic bronchoscopy	radiology	transthoracic biopsy
bronchial tree secretions, bacteriology and cytology	plain PA and lateral views	fine needle aspiration
bronchial biopsy	plain tomography	cutting needle biopsy
transbronchial biopsy	xerography	
transbronchial needle aspiration	computerized axial tomography	
selective bronchial brushing	digital subtraction imaging	
bronchoalveolar lavage		

Fig. 3.2 Diagnostic methods for the assessment of mass lesions.

the exact sites or appearances of discrete lesions may sometimes suggest an inital diagnosis, for example, an apical round shadow with a lucent rim of air indicates a mycetoma (Fig. 3.3) and lobar or segmental collapse suggests a proximal obstructing lesion such as a bronchial carcinoma, further tests to confirm or refute the suspicion will be necessary.

Fibreoptic Bronchoscopy

Endoscopic examination of the bronchial tree is vital in the assessment of a patient with a mass on the chest radiograph, regardless of site or whether the cause is suspected to be benign or malignant. Since many such patients may have resectable lesions, it is essential that bronchoscopists are trained in judging a patient's potential for surgical management.

Bronchial tree secretions

SPUTUM TRAP BACTERIOLOGY – Sputum trap culture of specimens obtained through the fibrescope suction channel are seldom diagnostic, because of contamination of the instrument by passage through nasal and pharyngeal secretions. The value of such material may be increased by putting the suction trap into the circuit after entry into the trachea. In a case where localized infection is suspected, a sterile, double lumen, plugged suction catheter or brush may be introduced and useful material aspirated from the suspected site by washing with 30-60ml of normal saline. Where patients are immunocompromised, culture onto special media at the time of endoscopy may be required to ensure growth of anaerobic bacteria.

In theory, the preservative 0.1% chlorocresol, present in the four per cent topical lignocaine solution used for

local anaesthesia of the larynx and vocal cords, may inhibit bacterial growth if it contaminates the trap, but in practice this is seldom a problem. The two per cent solution used for the bronchial tree itself contains no additives and does not affect sputum culture.

Tuberculous infection, which is suspected to have caused a mass, may also need to be diagnosed by lavage at fibreoptic bronchoscopy after conventional sputum microscopy and culture have failed. Where this infection is a possibility, it will always be necessary to notify the endoscopy unit's staff responsible for instrument cleaning and theatre sterility. Any risk to subsequent patients may be avoided by using the fibreoptic bronchoscope last on suspected tuberculous patients, sterilizing the instruments in glutaraldehyde for at least sixty minutes or by reserving a dedicated instrument for the purpose.

SPUTUM TRAP CYTOLOGY – Cytological examination of trap sputum should always be carried out in patients with mass lesions on the chest radiograph when malignancy is suspected. However, it must be strongly emphasized that its value depends on the interpretive skills of the cytologist. The investigation is seldom useful if an endobronchial tumour is visible for biopsy, but where tumours are either not visible or are peripherally sited, positive sputum cytology may contribute.

Experience has shown, however, that cytological cell type cannot be relied upon to give a definitive histological diagnosis of tumour type. Furthermore, potentially worrying and spurious positive sputum cytology may result either from patients with current or previous infection, particularly tuberculosis, or in those who have had a recent bronchoscopy. It is therefore unwise to decide on future management solely on the result of a single positive sputum cytology test. Needle biopsy or even immediate diagnostic thoracotomy may be considered necessary in cases where the only evidence for malignancy is derived from a cytological examination.

Bronchial biopsy

Histological material may readily be obtained from visible lesions in the bronchial tree. Care should be taken to send specimens for bacteriological culture and a smear for acid-fast bacilli should be performed if infection or tuberculosis is suspected. Biopsy of bronchial mucosa of a relatively normal appearance taken from the same anatomical site as suggested by the mass may sometimes yield positive results.

Where possible, biopsies should be taken from the edge of tumours; the presence of a diagnostically useful specimen being suggested by the ease of avulsion and by minimal bleeding from the site. A biopsy site which does not bleed suggests that inspissated mucus or necrotic, and therefore often not diagnostic, material has been taken. Care is needed at the edge of tumorous areas, from which histology is often reported as normal. If the tissue is difficult to 'bite off' and if it comes away in strips it is most likely that non-diagnostic material will have been obtained. This leads to frustrating delay and necessitates a further procedure.

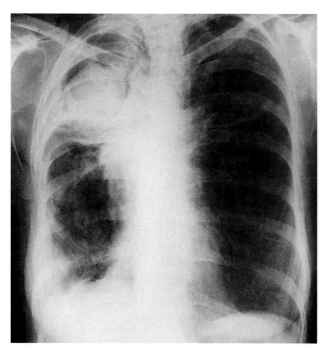

Fig. 3.3 Chest radiograph showing a typical mycetoma lying in an old right apical tuberculous cavity.

Transbronchial biopsy

Transbronchial forceps biopsy (see *Chapter 1*) may be valuable when diagnosing peripheral pulmonary lesions which are not endoscopically visible. The biopsy forceps should enter the lesion before specimens are taken and accurate placement may be ensured by screening under image intensification (Fig. 3.4). Eighty per cent of peripheral malignant tumours over 4cm in diameter and sixty per cent of lesions of 4cm diameter or less may be diagnosed in this way. An even greater yield is claimed by some workers using transbronchial curettage guided by prior bronchography, in which the diagnosis is made by cytological examination of curettings alone. These procedures are relatively safe and easy for a skilled operator, provided that bleeding disorders have been excluded, and may spare more invasive techniques. It must be emphasized that success rates vary widely and improve with the experience of the operator.

Transbronchial needle biopsy

Aspiration of cytological material by fine needle puncture of the bronchial wall under direct vision through the fibrescope may prove a useful means of obtaining diagnostic tissue, particularly from extrabronchial lesions (Fig. 3.5). The material which may not otherwise be available has been of value in staging bronchial neoplasms. This method enables biopsy through apparently normal looking bronchial mucosa and penetration into tumours outside the bronchial tree. The value of the technique remains to be fully assessed.

Selective bronchial brushing

Cytological material obtained from brushing in lobar, segmental or subsegmental bronchi, in which lesions appear on the chest radiograph but not endoscopically, may be a useful means of detecting malignancy, if present. However, its particular value is in patients with no visible chest radiographic abnormality who consistently have malignant cells in their sputum. For example, it is useful in the very early detection of bronchial carcinoma. The technique requires the separate brushing of all segmental bronchi to identify such an early malignant lesion.

The object of this chapter is to demonstrate the value of the fibreoptic bronchoscope in the investigation of disorders which may present with a discrete mass on chest radiography. Various types of radiological presentation are dealt with and accounts of these are illustrated by case histories and endoscopic findings.

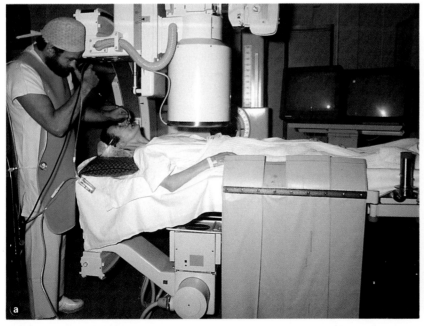
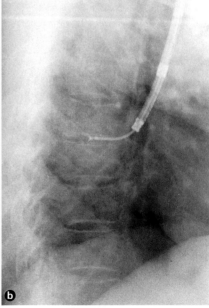

Fig. 3.4 Fluoroscopic image intensification. View (a) shows a patient undergoing transbronchial biopsy under 'C' arm fluoroscopic control. Note screens for direct viewing in the background. In view (b), fibreoptic biopsy forceps are shown located in an ill-defined squamous cell carcinoma in the apical segment of the right lower lobe.

LUNG FIELD, HILAR AND MEDIASTINAL LESIONS

These may be benign or malignant presenting as lung collapse or discrete shadows.

Primary Bronchial Carcinoma

The radiological appearance of a lung collapse may suggest a central obstructing lesion.

Case Study 1

A sixty-eight-year-old retired labourer known to have widespread osteoarthritis became bedridden with joint pains and had lost 13kg in weight over four weeks. On admission to hospital he was found to have right-sided chest pain, a persistent cough and he admitted to smoking thirty cigarettes per day. He was ill, wasted, anaemic and had finger-clubbing. A chest radiograph showed almost complete collapse

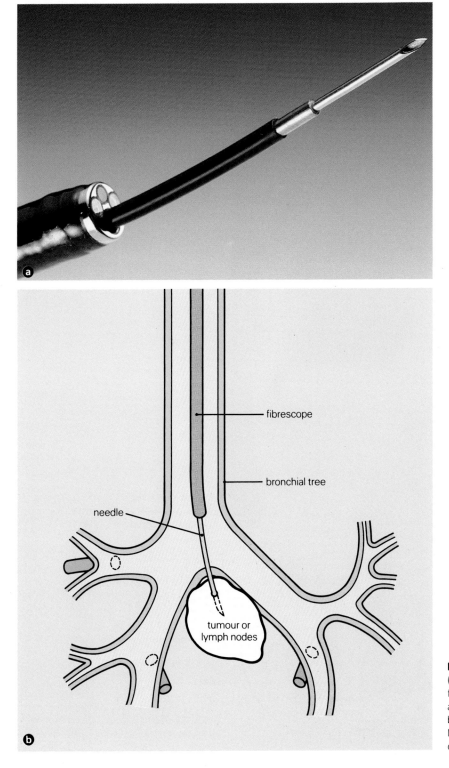

Fig. 3.5 Transbronchial needle biopsy. View (a) shows the retractable needle emerging from the fibrescope tip. The diagram (b) shows a transbronchial biopsy needle penetrating the bronchial wall and entering a mass of subcarinal lymph nodes or tumour. Photograph by courtesy of Key Med.

of the right lung (Fig. 3.6). Fibreoptic bronchoscopy revealed that the right main bronchus was narrowed by visible tumour at carinal level (Fig. 3.7) and occluded by extrinsic compression involving the posterior wall of the right main bronchus. These findings rendered him inoperable. Histology showed invasive squamous cell carcinoma and a bone scan confirmed bony metastases. The patient died a few weeks later.

Similarly the collapse of a single lobe may lead the endoscopist to expect a more peripheral bronchial lesion.

Case Study 2

A sixty-four-year-old woman developed a flu-like illness with left-sided pleuritic pain and a cough with purulent sputum. Two days later a severe pain developed in the mastoid region with subsequent right facial weakness. She felt lethargic and had noted a gruff voice over the previous two months. On examination, she had a puffy face with a right lower motor neuron facial weakness, but there were no other neurological signs. She had a 2cm diameter hard intracutaneous lump over her sacrum and there were signs of a left pleural effusion. A left hilar mass and evidence of left lower lobe collapse was confirmed by chest radiography (Fig. 3.8).

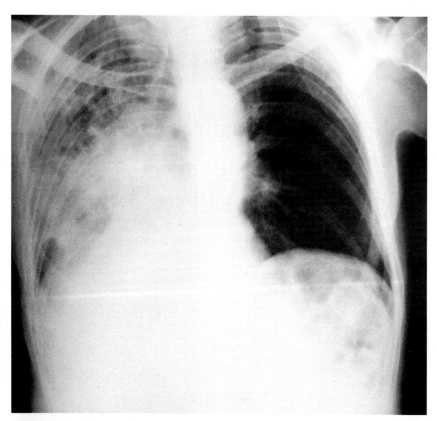

Fig. 3.6 Case study 1. Chest radiograph showing partial right lung collapse due to bronchial carcinoma.

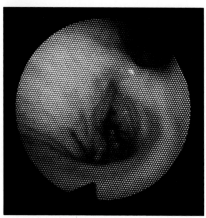
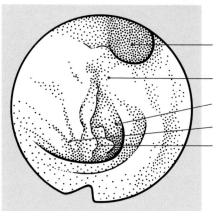

— orifice of left main bronchus

— carina

— orifice of right main bronchus

— posterior extrinsic compression

— tumour

Fig. 3.7 Case study 1. Squamous carcinoma and posterior extrinsic compression occluding the right main bronchus just below carinal level.

3.6

Fibreoptic bronchoscopy revealed that the left lower lobe bronchus was occluded by a tumour and that the upper lobe bronchus was narrowed by mucosal oedema and extrinsic compression (Fig. 3.9). Forceps and brush biopsy, sputum trap cytology and, later, rigid bronchoscopy were all negative for malignancy. The diagnosis of adenocarcinoma of uncertain origin was subsequently confirmed by Abrams' needle biopsy of the pleura and from excision of the cutaneous sacral nodule. No primary renal, breast or gynaecological source could be found and it was concluded that the lesion arose in the left lower lobe bronchus. A computerized tomographic (CT) scan of the head failed to reveal any intracranial cause for her facial palsy. This case illustrates that despite seeing a left lower lobe bronchial tumour it was not possible to confirm the diagnosis by biopsy via the endobronchial route, even by rigid bronchoscopy.

Primary carcinoma of the bronchus may occasionally arise at more than one site. *Case Study 3* illustrates such an occurrence and also shows how some peripheral tumours may be visible endoscopically whilst others may not.

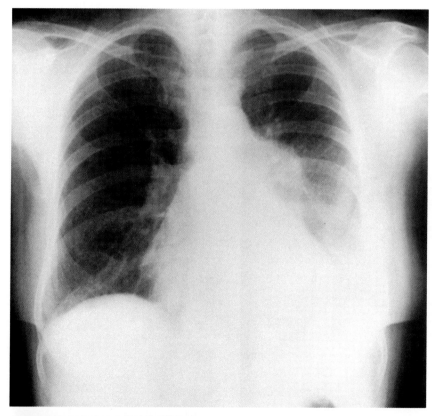

Fig. 3.8 Case study 2. Chest radiograph of a woman with primary adenocarcinoma causing a left hilar mass, left lower lobe collapse and left pleural effusion.

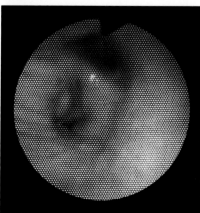
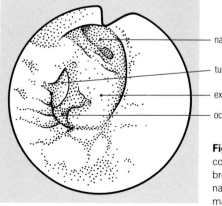

narrowed left upper lobe bronchus

tumour

extrinsic compression

occluded left lower lobe bronchus

Fig. 3.9 Case study 2. Tumour and extrinsic compression occluding the left lower lobe bronchus. The left upper lobe bronchus is also narrowed. Fibreoptic biopsy was negative for malignancy.

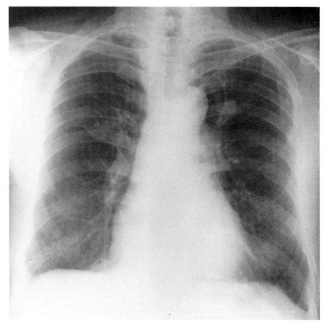

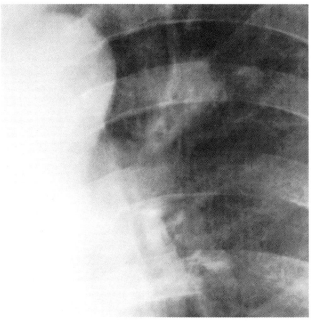

Fig. 3.10 Case study 3. Chest radiograph of a man with bilateral tumours. A soft irregular opacity above the right hilum and an ill-defined 2cm diameter opacity in the anterior segment of the left upper lobe (see enlargement) can be seen.

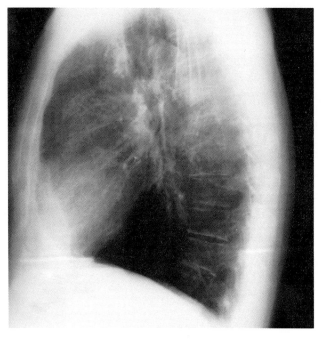

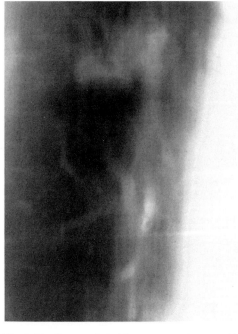

Fig. 3.11 Case study 3. Lateral chest radiograph of a man with bilateral tumours. A 2cm diameter soft shadow can be seen in the anterior segment of the left upper lobe.

Fig. 3.12 Case study 3. AP tomogram of a man with bilateral tumours. The tomogram confirms a lesion above the right hilum.

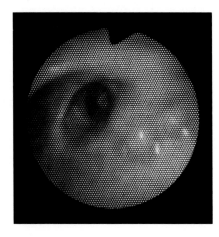

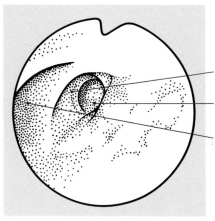

posterior segment of
right upper lobe bronchus

2mm diameter squamous cell carcinoma

orifice of apical segment of
right upper lobe

Fig. 3.13 Case study 3. Inoperable squamous carcinoma. A 2mm diameter tumour is present in the posterior segment of the right upper lobe bronchus.

Case Study 3

A seventy-six-year-old retired steel shearer gave up smoking twenty cigarettes per day because of cough and sputum and five months later developed a wheeze. The family practitioner carried out routine chest radiography at the local chest clinic and this showed an ill-defined 2cm diameter opacity in the anterior segment of the left upper lobe and a soft irregular opacity above the right hilum (Figs 3.10 and 3.11). The latter was confirmed on an AP tomogram (Fig. 3.12). He was a well looking man with moderately severe obstructive airways disease having a forced expiratory volume (FEV_1) of 1.3 litres and forced vital capacity (FVC) of 2.15 litres.

Fibreoptic bronchoscopy showed a 2mm diameter tumour in the posterior segment of the right upper lobe bronchus (Fig. 3.13) and biopsy confirmed a squamous carcinoma. There was no visible lesion in the left bronchial tree but a brush biopsy in the anterior segment of the left upper lobe revealed grade IV malignant squamous cells. Thus on radiological, endoscopic and cytological grounds there was strong presumptive evidence of bilateral, possibly primary, tumours. He was clearly inoperable and as he had few symptoms referrable to these lesions a conservative approach was adopted.

Peripheral lung field lesions may be beyond the range of vision of the fibreoptic bronchoscope and in these cases transbronchial biopsy may be of value in making the diagnosis.

Case Study 4

A fifty-year-old man was an asymptomatic close contact of a patient with sputum positive tuberculosis. Routine chest radiography (Fig. 3.14) showed a 3cm diameter shadow in the apical segment of the right lower lobe. Fibreoptic bronchoscopy revealed no endobronchial abnormality but histology of a transbronchial biopsy showed a keratinized squamous cell carcinoma. This was completely excised by right lower lobectomy and the patient remains alive and well ten years later.

Assessment for surgery

As a majority of patients with masses on chest radiography will present to physicians, the contribution of fibreoptic bronchoscopy and associated techniques to surgical assessment will be reviewed. The bronchoscopist should have all available radiological views to assess

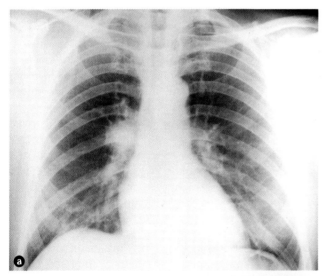

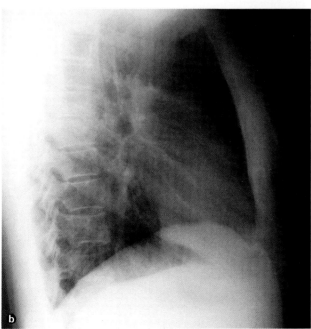

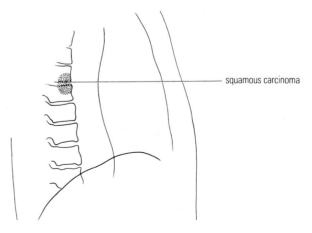

Fig. 3.14 Case study 4. Chest radiography of a man with squamous cell carcinoma. A 3cm diameter shadow is seen in the apical segment of the right lower lobe; PA (a) and lateral (b) views.

the site and relations of the lesion to be investigated, and knowledge of lung function, haematological and biochemical measurements, as well as a full clinical assessment of the patient. At the outset, the physician bronchoscopist must remember that usually he or she is not also a surgeon and that, whenever doubt exists, it will be necessary to discuss with, or refer to, a thoracic surgical colleague. It is essential that no patient, who might have an operable tumour, is denied the benefit of such opinion.

VOCAL CORDS – Endoscopic assessment of the bronchial tree commences in the larynx, whether or not a voice disturbance is present. After the vocal cords have been anaesthetized they are inspected for localized lesions such as polyps or tumours. These abnormalities are common and should not be biopsied at bronchoscopy, but left to an ear, nose and throat (ENT) surgeon. Vocal cord paralysis is usually detected in those undergoing local anaesthesia by reduction of movement, particularly adduction, when the patient is asked to say 'eeeee'. Most commonly, hoarseness due to paralysis of the left cord is due to involvement of the left recurrent laryngeal nerve, as it passes under the aorta, by carcinoma at the left hilum. This observation means that the patient has an inoperable growth.

Case Study 5
A seventy-one-year-old man presented with four months breathlessness, hoarseness and left-sided chest pain. The chest radiograph showed a pleural effusion, a mass at the left hilum and probable tumour in the subaortic fossa (Fig. 3.15). At fibreoptic bronchoscopy there was a left vocal cord paralysis (Fig 3.16). This proved to be due to a squamous cell carcinoma in the left upper lobe bronchus

Right vocal cord paralysis due to carcinoma is much less common and when present is due to a right lung tumour

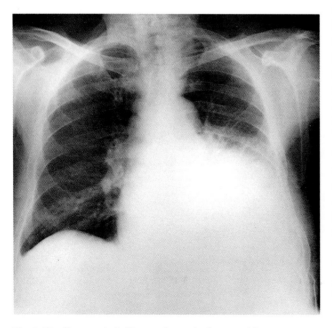

Fig. 3.15 Case study 5. Chest radiograph of a man with squamous cell carcinoma in the left upper lobe bronchus. Tumour at the left hilum with pleural effusion can be seen.

invading beyond the apical pleura to involve the recurrent laryngeal nerve as it hooks around the right subclavian artery. This also renders the patient inoperable.

Detection of obvious paralysis presents no problems, but when a patient with a hoarse voice and an obvious radiographic mass has no definite visible cord abnormality or asymmetry of movement on phonation at bronchoscopy, an assessment by an ENT surgeon is essential as more accurate information can be gained about vocal cord movement and associated structures at direct laryngoscopy.

TRACHEA – The best view of the trachea is obtained by looking along its entire length immediately after passing between the vocal cords (Fig. 3.17). Mobility of the tracheal wall may be assessed by watching movement, particularly of the non-cartilaginous posterior wall on coughing, deep inspiration and expiration. This may be lost if tumour either is sited on the mucosa or infiltrates it or is sited outside the tracheal wall (Fig. 3.18). From the same vantage point any deviation, narrowing or endotracheal tumour may be seen.

The siting of any abnormality and its relationship and distance from the larynx or carina should be described and measured at the nose with reference to the white 5cm marker rings on the fibrescope shaft. It is unwise to attempt to pass the fibrescope through tight tracheal stenoses even if it looks feasible for fear of causing bleeding and mucosal oedema and subsequently more severe obstruction. The same should apply to biopsy of such lesions as this procedure can be more safely carried out at rigid bronchoscopy. However, if the degree of narrowing is not great, assessment may be continued. If tumour is seen involving the lower end of the trachea and main bronchi, such patients are unlikely to be operable. However, primary tracheal carcinoma, unless obviously very extensive, requires an expert surgical assessment.

CARINA – The appearance and mobility of the carina are important factors in the assessment of lung cancer for operability. It should have a sharp clearly-defined appearance (Fig. 3.19), becoming increasingly biconcave on inspiration but with compression and deformity in an anterior to posterior axis on coughing. Inflammation or broadening with loss of mobility and definition may suggest enlarged metastatic subcarinal lymph nodes. Visible infiltration of the carina by obvious tumour on one or both sides may be seen. Finally the carina may be distended or squashed by extrinsic compression by enlarged extratracheal lymph nodes or tumour tissue. Inflammation and distortion are not necessarily adverse features but carinal broadening or infiltration by tumour may suggest inoperability.

Case Study 6
A sixty-six-year-old woman presented with a five-month history of wheeze, a cough productive of sputum and episodic right-sided chest pain. Her chest radiograph showed right lower lobe collapse. In addition, she had lost 6kg in weight in that period and had experienced recent upper abdominal pain. She had smoked thirty cigarettes per day for over forty years.

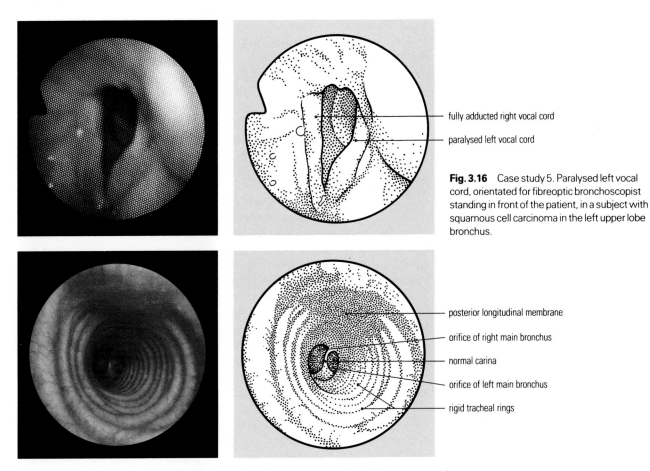

fully adducted right vocal cord

paralysed left vocal cord

Fig. 3.16 Case study 5. Paralysed left vocal cord, orientated for fibreoptic bronchoscopist standing in front of the patient, in a subject with squamous cell carcinoma in the left upper lobe bronchus.

posterior longitudinal membrane

orifice of right main bronchus

normal carina

orifice of left main bronchus

rigid tracheal rings

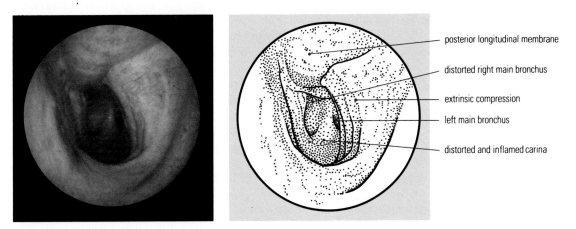

Fig. 3.17 A normal trachea from just inside the vocal cords, photographed via a rigid bronchoscope for clarity but orientated for a bronchoscopist standing in front of the subject. By courtesy of Dr P. Stradling.

posterior longitudinal membrane

distorted right main bronchus

extrinsic compression

left main bronchus

distorted and inflamed carina

Fig. 3.18 Extrinsic compression of the left lower trachea with carinal and right main bronchial distortion by tumour, photographed via a rigid bronchoscope for clarity but orientated for a bronchoscopist standing in front of the subject. By courtesy of Dr P. Stradling.

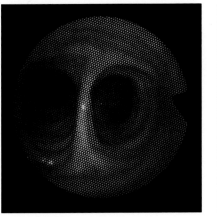

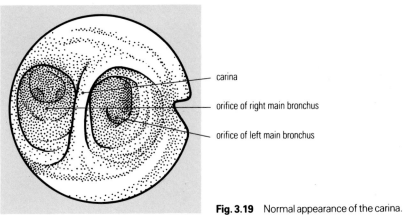

carina

orifice of right main bronchus

orifice of left main bronchus

Fig. 3.19 Normal appearance of the carina.

Examination showed signs of right lower lobe collapse; she had an enlarged liver and liver enzymes were elevated. Spirometry showed an FEV_1 of 0.75 litres and FVC of 1.2 litres suggesting severe airway obstruction. Fibreoptic bronchoscopy showed a normal upper airway but the carina was involved posteriorly by tumour, being immobile and infiltrated on both sides (Fig. 3.20). This together with poor lung function rendered her condition inoperable. The right upper lobe bronchus was occluded by tumour which also partially obstructed the main bronchus and this could not be entered. The histology of the biopsy specimen showed a small cell (oat cell) carcinoma. She was considered suitable for radiotherapy.

MAIN BRONCHI – The approximate siting of a tumour endobronchially can often be predicted from the chest radiograph. The expected normal side should be bronchoscoped first, inspecting each segmental and sub-segmental orifice sequentially. A meticulous search of all the visible airways provides good practice for less experienced operators, but most importantly excludes unexpected contralateral lesions which might complete-ly alter future management.

On the affected side, the precise siting of bronchial narrowing is noted, observations being made regarding inflammation, extrinsic compression and tumour. It is important to note whether tumour arises at or beyond the actual bronchial orifice and whether it appears to extend within the walls of the parent bronchus, either above, below or encircling it. Additionally, mobility of the bronchial walls with respiratory movements is noted. It should be remembered that this particular aspect of assessment cannot be relied upon with the flexible fibrescope and there is little doubt that mobility or lack of it in the bronchial tree is better assessed with a rigid bronchoscope. This is of course a part of the surgeon's pre-operative inspection.

Once the tumour has been inspected, it is then necessary to attempt to pass beyond it to the remainder of the bronchial tree on the affected side, noting distal involvement if any. A normal bronchial tree beyond a tumour may suggest to the surgeon that the unaffected lung may be conserved. Caution is needed with tumours that appear to occlude the bronchus as some of these may be polypoid, truly arising in the bronchial tree below their apparent siting. This would suggest a more favour-able surgical prognosis and may enable a resection lower in the bronchial tree.

Tumour either involving the high medial wall of the right main bronchus or situated at the carina may be inoperable, but mucosal lesions on the lateral wall of the trachea above carinal level may be amenable to resec-tion. Tumour which involves the bronchial orifice of the right upper lobe but goes no higher may require pneumonectomy. Lesions of the middle lobe orifice may need middle and perhaps lower lobectomy, depending on the degree of intra- and extrabronchial involvement. Involvement of the lower lobe orifice or bronchus alone is most favourable for resection. Commencement of tumours distally in either lobar or segmental bronchi beyond the site of origin is also a favourable factor.

Anatomical considerations on the left side are similar. Tumours near to the carina are often inoperable, but again unless there is obvious involvement of the carina the surgeon must be asked to decide what is surgically feasible. At the completion of visual assessment, biopsies are taken, making sure to identify individually those from different sites and sides. If biopsy is not left until the end, visual assessment may be hindered and time wasted by fogging of the lens with blood, which may readily reduce viewing with the fibreoptic instrument.

Finally, before leaving the bronchial tree, the operator should check that bleeding is not excessive and take a further and final view of the upper tracheobronchial tree, larynx and pharynx whilst the instrument is slowly withdrawn.

Physicians carrying out bronchoscopy should be ever mindful that they cannot themselves see outside the bronchial tree and are not going to be involved in eventual surgery. They should not therefore get involved in the assessment of where resection lines may or may not be, nor should they make any but the most general of surgical assessments. It cannot be repeated often enough that, where there remains doubt about operability, a thoracic surgical opinion should be sought. For these reasons it is mandatory that all fibreoptic bronchoscopists should have some experience of rigid bronchoscopy and spend some time in training on a thoracic surgical unit.

Secondary Bronchial Carcinoma
Uncommonly, endobronchial tumours will be metastases from a distant site. Breast and renal carcinoma may more often behave in this way but most malignancies may metastasize to the lungs. There are no characteristic appearances of such tumours.

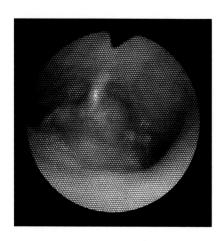
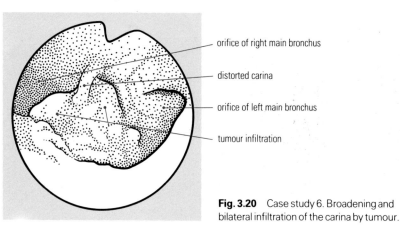

orifice of right main bronchus

distorted carina

orifice of left main bronchus

tumour infiltration

Fig. 3.20 Case study 6. Broadening and bilateral infiltration of the carina by tumour.

Case Study 7

A forty-year-old woman had a nephrectomy for a papillary adenocarcinoma of the kidney and remained well over the next ten years. At age fifty-four, a number of discrete shadows presumed to be metastases appeared in the left lung (Fig. 3.21) and she was started on high dose progestogen (medroxyprogesterone acetate) therapy. The size of these lesions did not change over three years, chlorambucil therapy being started with benefit when they did enlarge.

Two years later she presented with an eight-month history of a dry cough, haemoptysis and breathlessness with an irregular pulmonary shadow in the left mid-zone (Fig. 3.22). Anaemia (Hb 8.5g/dl) and thrombocytopenia (35,000 platelets/dl) were ascribed to chlorambucil which was stopped after bony metastases had been excluded by a bone scan. Fibreoptic bronchoscopy showed tumour recurrence in the lingula (Figs 3.23 and 3.24) and lateral segment of the left lower lobe and histological examination of the biopsy showed an adenocarcinoma compatible with a renal origin. She was treated with further chemotherapy but gradually deteriorated and died fourteen months later.

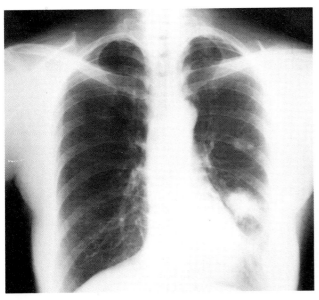

Fig. 3.21 Case study 7. Chest radiograph of a woman with renal adenocarcinoma metastases in the left lung. A number of discrete shadows are visible.

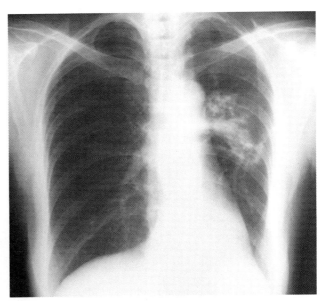

Fig. 3.22 Case study 7. Chest radiograph showing post-chemotherapy recurrence of metastases in the left mid-zone.

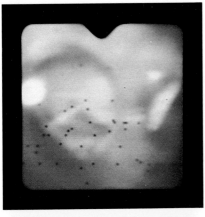

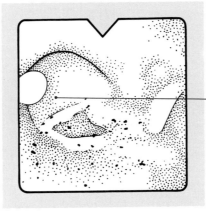

tumour in lingular bronchus

Fig. 3.23 Case study 7. Post-chemotherapy recurrence of lingular endobronchial renal metastasis.

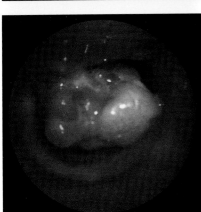

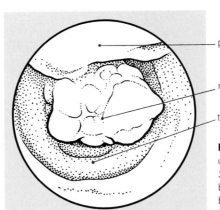

posterior tracheal membrane

renal adenocarcinoma metastasis

tracheal lumen

Fig. 3.24 Appearance of tracheal renal carcinoma metastasis, similar to that of *Case Study 7*, photographed through a rigid bronchoscope for clarity, but orientated for a bronchoscopist standing in front of the patient. By courtesy of Dr P. Stradling.

3.13

Case Study 8

A fifty-five-year-old woman underwent a simple mastectomy for carcinoma of the breast, developing a scar recurrence at one year and a hoarse voice after a further year. An assessment by an ENT surgeon showed a complete left vocal cord paralysis and chest radiography and CT scanning revealed a large left paratracheal mass extending from the superior mediastinum into the apex (Figs 3.25 and 3.26). Fibreoptic bronchoscopy showed inflammation and infiltrated mucosa in the left upper lobe (Fig. 3.27). Histological examination of a biopsy of the lesion revealed poorly differentiated squamous malignant cells which strongly resembled the polygonal cells of squamoid appearance seen in the original mastectomy specimen. Following the commencement of chemotherapy her upper mediastinal mass regressed.

In this case, bronchoscopy was helpful in resolving the problem presented by a mediastinal mass.

Bronchial Adenomas

In the past, these tumours have been regarded as benign but it is now known that they may eventually behave in a malignant fashion, spreading locally and to distant sites. Ninety per cent are of the carcinoid type, the remainder being cylindromas or the rarer mucoepidermoid or

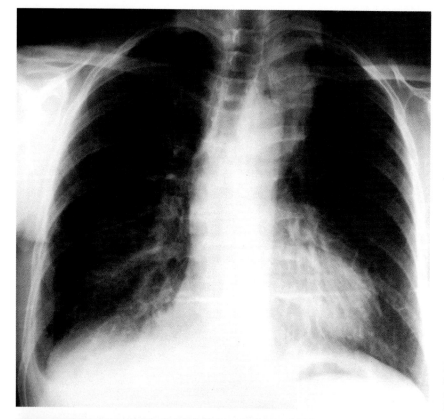

Fig. 3.25 Case study 8. Chest radiograph showing metastatic breast carcinoma extending from the mediastinum into the left lung apex.

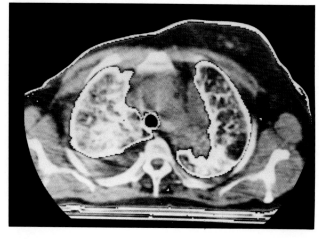

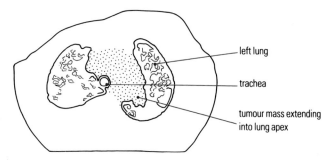

left lung

trachea

tumour mass extending into lung apex

Fig. 3.26 Case study 8. Computerized axial tomogram showing a mediastinal mass extending into the apex of the left lung.

mixed tumour. *Case Study 9* illustrates the presentation of such a tumour.

Case Study 9

An eighteen-year-old girl had a prominence in the right hilar region shown on a routine chest radiograph (Fig. 3.28). She was investigated with negative results but treated for tuberculosis three years later. After another year she became breathless, wheezed and had a dry cough with retrosternal pain on exercise. The chest radiograph abnormality persisted and was presumed to be due to sarcoidosis; she was treated with corticosteroids for her symptoms. Four years later she was investigated at a different centre, this time with increased breathlessness and backache and on examination was found to have abnormal facial hair, dyspnoea at rest and signs of mitral incompetence. Echocardiography showed a thickened interventricular septum, reduced left ventricular cavity size and systolic posterior cusp prolapse. Pulmonary function tests showed a marked restrictive defect with a reduction of gas transfer coefficient to seventy-three per cent of predicted. The chest radiograph (Fig. 3.29) showed a prominent right hilum and diffuse bilateral pulmonary shadowing.

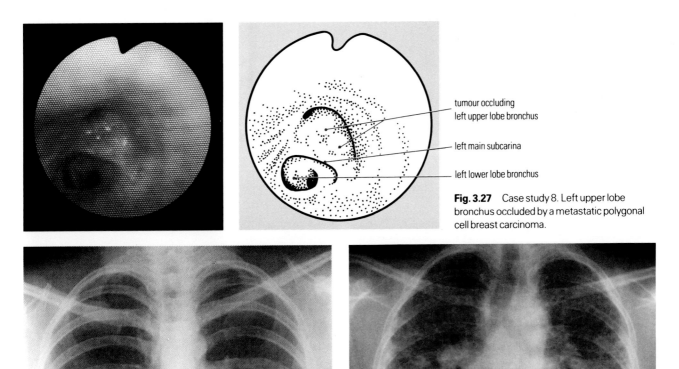

tumour occluding
left upper lobe bronchus

left main subcarina

left lower lobe bronchus

Fig. 3.27 Case study 8. Left upper lobe bronchus occluded by a metastatic polygonal cell breast carcinoma.

Fig. 3.28 Case study 9. Chest radiograph of a girl with a bronchial adenoma of the carcinoid type. The tumour is seen as a prominence in the right hilar region.

Fig. 3.29 Case study 9. Chest radiograph five years later shows diffuse pulmonary shadowing as well as a prominent right hilum.

Fibreoptic bronchoscopy revealed a tumour in the apical segment of the right lower lobe bronchus (Fig. 3.30) and bronchial and transbronchial biopsies were performed (Fig. 3.31). Both produced tissue suggestive of a carcinoid tumour infiltrating the lung. Lumbar spinal films showed partial collapse of the first lumbar vertebra (Fig. 3.32). The diagnosis was finally confirmed by finding a whole blood 5-hydroxytryptamine level of 705ng/ml (normal range, 100 – 250ng/ml) and an elevated twenty-four-hour urinary 5-hydroxy-indoleacetic acid level of 56mg (normal range, <20mg).

Lesions revealed by radiography may occur in the lung, at the hilum or in the mediastinum in patients with lymphomas or Hodgkin's disease. Either the presenting tumour is in the chest or recurrences manifest themselves in this way. Whichever is the case, a positive diagnosis is necessary to plan management.

Case Study 10

A twenty-four-year-old woman developed bilateral, non-tender fixed

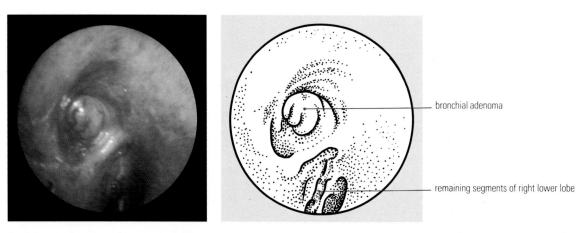

Fig. 3.30 Typical appearance of a bronchial adenoma in the right apical lower lobe bronchus similar to that of *Case Study 9*, photographed through a rigid bronchoscope for clarity, but orientated for a broncho-scopist standing in front of the patient. By courtesy of Dr P. Stradling.

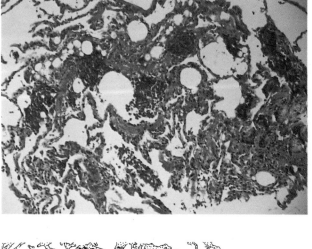

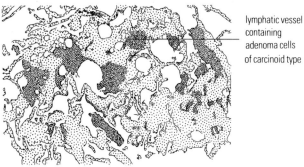

Fig. 3.31 Section of transbronchial biopsy of lung tissue showing lymphatic infiltration by carcinoid tumour. (H and E).

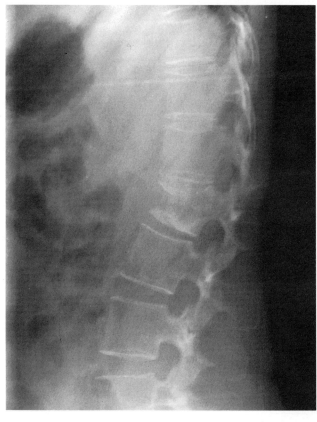

Fig. 3.32 Case study 9. Lateral radiograph showing erosion and collapse of the first lumbar vertebra by metastatic carcinoid tumour.

lymph nodes in the supraclavicular region and anterior cervical chains and subsequently developed a discharging sinus on one side. She had a haemoglobin of 8.5g/dl, a white cell count of 22,000/mm³, with ninety per cent neutrophils, and an ESR of 90mm/hr. The chest radiograph showed right paratracheal glands and a prominent right hilum (Fig. 3.33). Cervical lymph node biopsy produced tissue containing clusters of foreign-body type giant cells, but no granulomata or fungi were seen and Ziehl-Nielsen staining for alcohol and acid-fast bacilli (AAFB) was negative. Occasional histocytic cells with eosinophilic nuclei morphologically reminiscent of Hodgkin's-type Reed-Sternberg cells were seen. However, lacking a definite diagnosis, she was started on anti-tuberculous treatment with rifampicin and isoniazid without benefit.

Twelve months after first presentation, a further lymph node biopsy was performed at a second centre and this contained tissue which was typical of nodular sclerosing Hodgkin's disease and she was graded IIIB. She was treated with mustine, vincristine, procarbazine and

prednisolone but defaulted after an incomplete course. After a further increase in lymph node size she was persuaded to return for further chemotherapy and she complied with a full course. On this occasion her cervical and mediastinal nodes regressed with healing of the sinuses.

One year later she again developed right hilar and pulmonary shadowing (Fig. 3.34). She underwent fibreoptic bronchoscopy and this showed a tumour occluding the lateral segment of the middle lobe (Fig. 3.35). Histology revealed epithelioid granulomata and features typical of Hodgkin's lymphoma; Ziehl-Nielsen staining for AAFB was negative but six weeks later the bronchoscopy trap sputum grew *Mycobacterium tuberculosis*. She was commenced on quadruple therapy with rifampicin, isoniazid, pyrazinamide and ethambutol and subsequently a further course of chemotherapy for her recurrent Hodgkin's disease was given. This unusual case illustrates the simultaneous occurrence of double pathology in the endobronchial tumour.

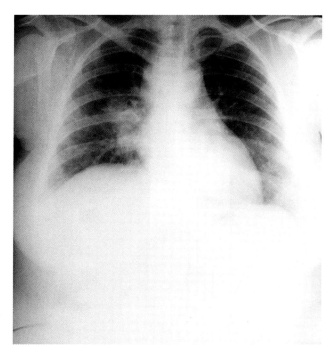

Fig. 3.33 Case study 10. Chest radiograph showing right hilar and mediastinal involvement by Hodgkin's disease.

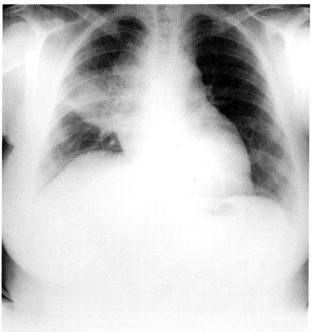

Fig. 3.34 Case study 10. Chest radiograph two years later showing hilar and pulmonary recurrence of Hodgkin's disease.

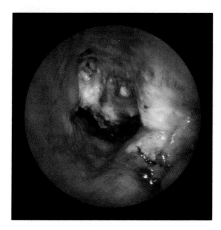

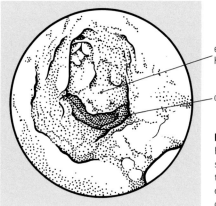

endobronchial tumour of Hodgkin's disease

orifice of left main bronchus

Fig. 3.35 Typical appearance of endobronchial Hodgkin's disease in the left lower bronchus, similar to that in *Case Study 10*, photographed through the rigid bronchoscope for clarity, but orientated for a bronchoscopist standing in front of the patient. By courtesy of Dr P. Stradling.

Benign endobronchial lesions may be found particularly in patients undergoing bronchoscopy for other lesions.

Case Study 11

A seventy-four-year-old man who smoked ten cigarettes per day went to his general practitioner with cough, wheeze and breathlessness and chest radiography was carried out. This showed an irregular lobulated 3×5cm shadow in the apical segment of the right lower lobe and a pneumonic area in the middle lobe (Fig. 3.36). Clinical examination revealed a well-looking man and there were no physical signs. Fibreoptic bronchoscopy was normal except for a 4mm diameter sessile polypoid lesion in the right upper lobe bronchus (Fig. 3.37), which on biopsy proved to be *bronchopathia osteoplastica*, a rare disorder, commoner in elderly men, characterized by cartilaginous and calcified plaques projecting into the bronchial lumen but not occluding it. As no diagnosis for the obvious radiological mass was available, an aspiration needle biopsy was performed which revealed grade IV malignant squamous cells. In view of a combination of age and poor lung function, he received localized palliative radiotherapy for the carcinoma.

Diagnosis of Bacterial and Fungal Infections

Bacterial and fungal infections may present as mass lesions on the chest radiograph. Such diagnoses should always be suspected, where relevant, and aspirated material sent for culture. The following case studies illustrate the value of this practice in acute bacterial and tuberculous infections as well as in fungal infections.

Case Study 12

A fifty-three-year-old non-smoking man developed a right basal pneumonia which resolved on antibiotics. Subsequently a thin-walled cystic lesion appeared at the same site in the posterior basal segment of the right lower lobe (Fig. 3.38) and because of it he was followed in a chest clinic over the next ten years. He remained asymptomatic until aged sixty-three when he developed mumps which was immediately followed by a dry cough, pleuritic right basal chest pain, night sweats and fever.

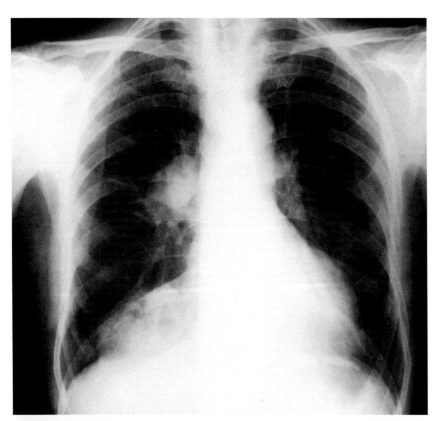

Fig. 3.36 Case study 11. Chest radiograph of squamous cell carcinoma of the apical segment of the right lower lobe.

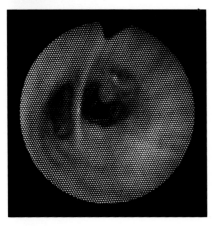

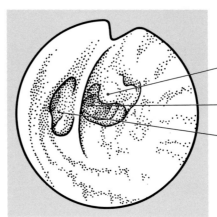

polypoid sessile tumour of *bronchopathia osteoplastica*

right upper lobe bronchus

orifice of apical segment of right upper lobe

Fig. 3.37 Case study 11. Polypoid *bronchopathia osteoplastica*. This was an incidental finding in the right upper lobe bronchus during investigation for another lesion.

On admission, he had a low grade fluctuating temperature of less than 38 °C and a right pleural effusion. The haemoglobin was 14.5g/dl, white cell count 13,700/mm³ and ESR 108mm/hr with a serum albumen of 24g/litre. The chest radiograph showed a 9cm diameter thin-walled abscess containing a fluid level at the same site as the original cyst (Fig. 3.39). Needle aspiration was not performed because

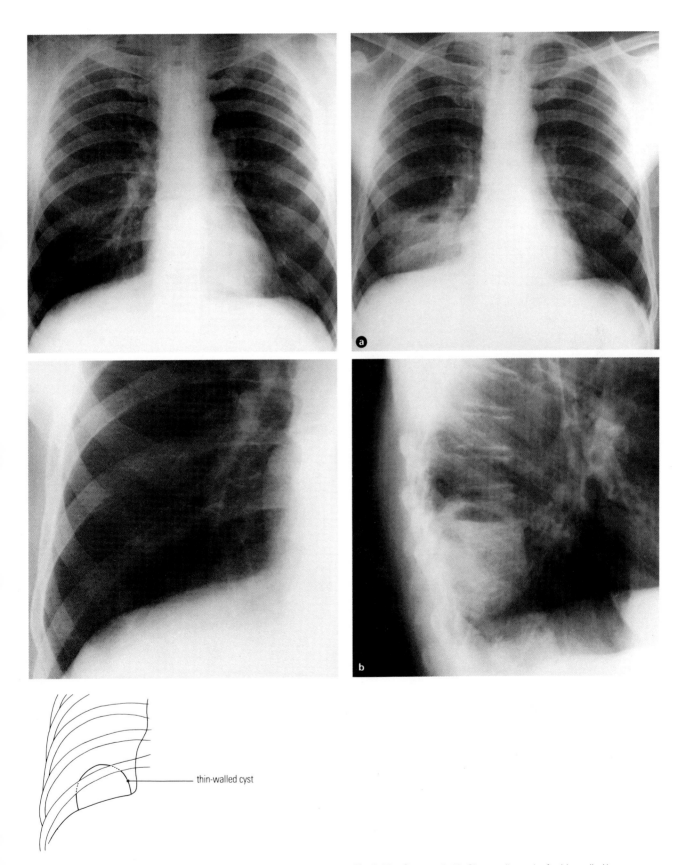

thin-walled cyst

Fig. 3.38 Case study 12. Full chest radiograph and enlargement showing a thin-walled post-pneumonic cyst above the right diaphragm.

Fig. 3.39 Case study 12. Chest radiograph of a thin-walled lung abscess. The abscess developed from a lung cyst in the posterior basal segment of the right lower lobe; P A (a) and lateral (b) views.

of the danger of producing an empyema. *Staphylococcus aureus* was grown from one set of blood cultures and treatment was commenced with parenteral sodium fucidate and erythromycin. Fibreoptic bronchoscopy was performed and showed narrowing of the posterior segmental bronchus of the right lower lobe by inflamed bronchial mucosa and extrinsic compression (Fig. 3.40). Pus from this bronchus grew *Haemophilus influenzae*.

It was decided to resect the lesion after an antibiotic course had failed to clear it and a right lower lobectomy was performed. Culture from the abscess cavity also produced *Haemophilus influenzae*. This patient, who made a complete and rapid recovery, illustrates how the fibreoptic bronchoscope may be used to obtain secretions for culture.

Case Study 13
A seventy-six-year-old man first developed bilateral apical disease on his chest radiograph twenty-one years previously. This was due to *Mycobacterium tuberculosis* and he was treated with three months streptomycin and a total of two years isoniazid and para-amino salicylic acid with successful resolution. Five years later chest radiography suggested recurrence, but this proved to be due to *M. xenopii* which was treated with a prolonged course of streptomycin, rifampicin and prothionamide and again the condition became quiescent.

Sixteen years after his first presentation he lost 13kg in weight, had increased cough and sputum and was noted to consume one bottle of whisky per week. A chest radiograph (Fig. 3.41) showed old apical fibrosis and a thick-walled cavity on the left side. Sputum culture for acid-fast bacilli failed to grow and therefore fibreoptic bronchoscopy was performed. Endoscopic appearances were normal but sputum trap culture grew *M. xenopii* eight weeks later and he was restarted on treatment.

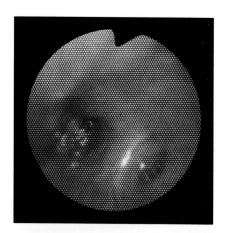

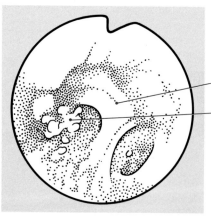

inflamed mucosa

pus exuding from posterior segment of right lower lobe

Fig. 3.40 Case study 12. Thin-walled lung abscess. Inflamed mucosa with pus exuding from the lateral segment of the right lower lobe can be seen. Note also bronchial narrowing by extrinsic compression due to the extrabronchial abscess.

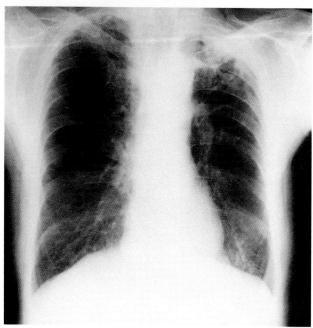

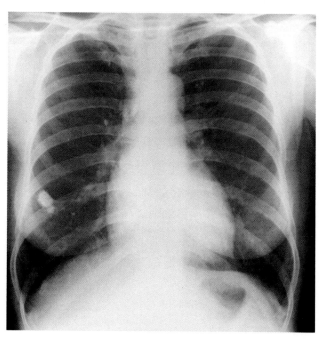

Fig. 3.41 Case study 13. Chest radiograph of a man with bilateral apical disease. There is left apical cavitation, lavaged secretions from which grew *Mycobacterium xenopii*.

Fig. 3.42 Case study 14. Chest radiograph showing shadows due to calcified tuberculosis.

Case Study 14

A fifty-five-year-old woman presented with right-sided chest discomfort and haemoptysis. She smoked twenty cigarettes per day and there was a history of treated tuberculosis twenty years previously. Her chest radiograph showed calcified intrapulmonary shadows characteristic of old healed disease (Fig. 3.42). Fibreoptic bronchoscopy showed a bronchostenosis 2cm from the origin of the middle lobe bronchus (Fig. 3.43). Biopsy produced fibrotic tissue only but the lesion was clearly the consequence of her old tuberculous infection. Mycobacterial culture was subsequently negative. There was no evidence of malignancy and her haemoptysis was assumed to result from her old healed tuberculosis.

Case Study 15

A fifty-eight-year-old man had a long history of winter bronchitis, a record of minor haemoptyses and had had several exacerbations of his airflow obstruction over eight months. He developed left-sided chest pain and lost 5kg over a six-month period. Forty years previously he had had pulmonary tuberculosis which responded to six months bed rest. Clinically he had a flat immobile left upper chest and percussion dullness at the right lung base. A chest radiograph showed old fibrocalcific shadowing, greater at the left apex, and a collapsed right lower lobe (Fig. 3.44). His sputum grew *M. tuberculosis* on culture and he was admitted for commencement of a nine-month course of rifampicin, isoniazid and ethambutol.

Routine investigations showed haemoglobin of 15.0g/dl, white cell count 11,500/mm^3, of which twenty-five per cent were eosinophils, and the ESR was 11mm/hr. *Aspergillus* precipitins were detected in the blood and skin prick testing with *Aspergillus fumigatus* antigen showed a skin reaction at fifteen minutes, but there was no delayed weal. Fibreoptic bronchoscopy showed occlusion of the right lower lobe bronchus by a rubbery grey plug which was extracted using biopsy forceps through the fibrescope (Fig. 3.45). Subsequently a radiograph

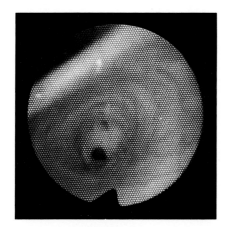
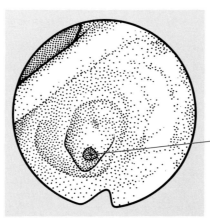

'pin-hole' stenosis in right middle lobe bronchus

Fig. 3.43 Case study 14. A narrow fibrous bronchostenosis in the middle lobe bronchus, revealed incidentally at fibreoptic bronchoscopy for haemoptysis.

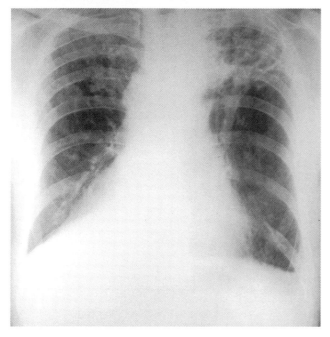

Fig. 3.44 Case study 15. Chest radiograph of a man with bronchopulmonary aspergillosis. Old healed apical tuberculosis and right lower lobe collapse can be seen.

Fig. 3.45 Case study 15. Rubbery grey bronchial plugs typical of bronchopulmonary aspergillosis. These were seen in the right lower lobe bronchus and extracted at fibreoptic bronchoscopy. By courtesy of Dr R. K. Knight.

showed the lobe to be re-expanded (Fig. 3.46) and later *Aspergillus fumigatus* was grown from the plugs. This patient had allergic bronchopulmonary aspergillosis and was subsequently treated with prednisolone without further recurrence.

Case Study 16

A sixty-three-year-old man presented with haemoptysis. His chest radiograph showed a 4cm diameter solid round shadow, contained in a cavity of the right upper lobe and surrounded by a narrow lucent crescent of air. This appearance was thought to be typical of a mycetoma. He underwent fibreoptic bronchoscopy, during which the cavity was entered via the posterior segmental bronchus of the right upper lobe. Figure 3.47 shows the mycetoma photographed in its cavity. Culture and histology subsequently confirmed *Aspergillus fumigatus* as the infecting fungus. Entry to such a cavity at endoscopy is rarely achieved.

CONCLUSION

The fibreoptic bronchoscope is a versatile instrument, which is essential for practitioners seeing patients with mass lesions on the chest radiograph. It gives autonomy to thoracic physicians, some of whom may be distant from major centres and without a thoracic surgeon near by. Over more than a decade it has completely changed the prospect for a new generation wishing to investigate their own patients, particularly subjects suspected of having endobronchial neoplasia or infection or those with parenchymal disease. Being relatively safe, easy and acceptable without general anaesthesia, fibrescopy of unfit or even moribund patients is now more than ever possible. It is facile to argue that the fibreoptic bronchoscope is superior to the rigid instrument in the investigation of mass lesions, but rather that the techniques should be seen as complimentary. For this reason basic training with both types of instrument is desirable.

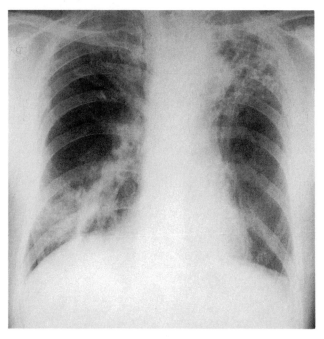

Fig. 3.46 Case study 15. Chest radiograph showing re-expansion of the right lower lobe two days after bronchoscopy and commencement of prednisolone treatment in a man with bronchopulmonary aspergillosis.

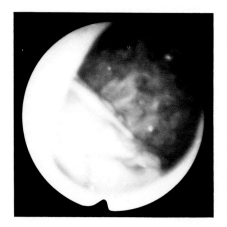

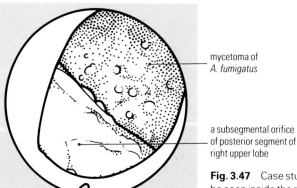

mycetoma of
A. fumigatus

a subsegmental orifice
of posterior segment of
right upper lobe

Fig. 3.47 Case study 16. A mycetoma can be seen inside the right upper lobe cavity. By courtesy of Dr R.K. Knight.

REFERENCES

Credle WF, Smiddy JF, Elliot RC (1974). Complications of fiberoptic bronchoscopy. *American Review of Respiratory Disease,* **109,** 67-72.

Ikeda S (1974). Atlas of Flexible Bronchofiberscopy. Igaku Shoin Ltd, Tokyo: 230pp.

Knight RK, Clarke SW (1979). An analysis of the first 300 fibreoptic bronchoscopies at the Brompton Hospital. *British Journal of Diseases of the Chest,* **73,** 113-120.

Martini N, McCormick PM (1978). Assessment of endoscopically visible bronchial carcinomas. *Chest,* **73,** 718-720.

Mitchell DM, Emerson CJ, Collins JV, Stableforth DE (1981). Transbronchial lung biopsy with the fibreoptic bronchoscope: analysis of results in 433 patients. *British Journal of Diseases of the Chest,* **75,** 258-262.

Sackman MA (1975). Bronchofiberscopy: state of the art. *American Review of Respiratory Disease,* **111,** 62-88.

Stableforth DE, Clarke SW (1976). Flexible fibreoptic bronchoscopy including a review of currently available equipment. *British Journal of Clinical Equipment,* **1,** 172.

4. Investigation of the Diffuse Lesion on Chest Radiography

R. M. du Bois MA MD MRCP

INTRODUCTION

The assessment of patients with diffuse lung disorders requires a methodical approach because of the vast number of causes of an interstitial pattern on chest radiography (Fig. 4.1). A detailed history, particularly past history, occupational history, hobbies, pets and drug history is essential. Patterns of radiological abnormality and lung function testing may be helpful, and certain immunological investigations will confirm some diagnoses which have been suspected after first assessment. However, despite thorough investigations some

Causes of Interstitial Lung Disease

immunological causes: collagen-vascular diseases

cryptogenic fibrosing alveolitis

extrinsic allergic alveolitis

pulmonary eosinophilia

granulomatous disorders

occupational causes: e.g. asbestosis, silicosis, siderosis, talcosis

infection: miliary tuberculosis

fungal infection, e.g. candidiasis

protozoan infection, e.g. *Pneumocystis*

viral infection, e.g. cytomegalovirus

drugs: e.g. amiodarone, cytotoxic drugs, Paraquat

rare causes: haemosiderosis

eosinophilic granuloma

alveolar proteinosis

neoplasia: multiple secondary deposits

alveolar cell carcinoma

lymphangitis carcinomatosis

leukaemia, lymphoma

Fig. 4.1 The causes of interstitial lung disease.

diagnoses will remain elusive without obtaining tissue and, until recently, open lung biopsy has been the only technique which has produced definitive samples. Use of the fibreoptic bronchoscope over the last ten to fifteen years has improved the diagnostic yield without recourse to open biopsy. In addition, techniques such as transbronchial biopsy, bronchoalveolar lavage and selective brush sampling have allowed a clear diagnosis to be made in many patients who would otherwise have been subjected to the more invasive biopsy procedure.

Transbronchial Biopsy

Despite initial worries that sample size would be inadequate for proper analysis, a sufficiently large biopsy can be obtained to enable competent pathologists to make unequivocal diagnoses in many conditions (e.g. sarcoidosis, tuberculosis and alveolar proteinosis). Alternatively, where there is considerable interstitial fibrosis (e.g. in fibrosing alveolitis) it becomes more difficult to obtain samples because the forceps cannot take an adequate 'bite' of the tissue. In this situation, there is no alternative but to perform open lung biopsy to confirm the diagnosis.

Samples obtained by transbronchial biopsy are stained routinely with haematoxylin and eosin (H and E) but, when indicated, additional information may be obtained by the use of special stains (e.g. to identify collagen or infective agents). The diagnostic yield may be improved by obtaining at least six samples at one session, which also enables material to be distributed to all the appropriate departments. This is particularly important in patients with infection as a consequence of immunosuppression. Samples should also be stored for electron microscopic analysis and can thus be reviewed, if necessary, at a future date.

Transbronchial biopsy may be hazardous in several situations; the most obvious of these is where platelet numbers are low or clotting is abnormal. Additionally, in certain situations, haemostasis is suboptimal even when platelet numbers and clotting are normal. In uraemia, for example, bleeding time may be prolonged despite other normal measurements; thus it is advisable to perform bleeding times wherever there is any doubt. If a transbronchial biopsy is thought to be essential despite suboptimal haemostasis, platelet and fresh frozen plasma cover may make the procedure acceptably safe; however, it is clearly not without risk under such circumstances.

Bronchoalveolar Lavage

Since the technique was first performed approximately ten years ago, bronchoalveolar lavage has become widely used for sampling larger volumes of lung than can be obtained by biopsy. Although the technique is used as a research tool in several institutes, it can be diagnostic in the investigation of some disorders such as haemosiderosis or alveolar proteinosis. The products of lung lavage can be broadly subdivided into the cellular and supernatant components.

Differential cell counts and electron microscopic examination of individual cells provide valuable informa-

tion about lung disease. Furthermore, new techniques (e.g. the use of monoclonal antibodies directed against cell surface antigens and microprobe analysis of inorganic inclusion particles) have expanded our knowledge of certain disease processes.

The supernatant is rich in substances which have been produced within the lung or which have diffused into the lung as part of an exudative process. Perhaps one of the most interesting areas of study on lung lavage supernatants is the assessment of proteases and their inhibitors. New inhibitors are currently being identified and new proteinases are also being found, and it seems likely that future studies of lavage material will help elucidate the pathogenesis of lung injury in the interstitial disorders.

Lung lavage of patients with interstitial lung diseases may be useful in three situations: (i) diagnostic lavage, for example in patients with haemosiderosis, alveolar proteinosis or eosinophilic granuloma; (ii) where additional information may help diagnosis, for example differential cell counts, or the use of monoclonal antibody techniques to identify viral particles in the lavage cell population obtained from patients with cytomegalovirus (CMV) infection; (iii) in the assessment of disease 'activity', progress and response to treatment. The latter group is more controversial and, until parameters are clearly defined, this must remain a research investigation rather than clinical practice.

It cannot be overemphasized that lung lavage must not be carried out as a routine in every district general hospital. Techniques of lavage vary throughout the world, and there is no consensus on the most ideal lavage procedure. Perhaps of more importance, is the handling of the samples and the processing of the cells to provide cytocentrifuge preparations which can then be stained and counted. If the material is not handled quickly and expertly erroneous results will be obtained; at best this will be misleading and at worst could result in serious errors of diagnosis. It is the author's view that, at present, bronchoalveolar lavage should be restricted to those centres which have expertise in it. It is not a liquid lung biopsy and should be regarded as a technique yielding additional data when making a diagnosis. Only in special circumstances does it provide unequivocal answers.

Selective Brush Sampling

This is the least useful technique employed in diagnosis of the interstitial lung disorders. It can be of value in the diagnosis of infection, particularly when a double lumen plugged catheter is used, and also in the diagnosis of diffuse pulmonary malignancies. In lung disorders resulting from occupational exposure, or diseases with an immunological basis, brush samples are of little value.

The causes of interstitial problems can be subdivided into five main groups: (i) immunological disease; (ii) occupational disorders; (iii) infection; (iv) drug-induced disease; (v) rarer lung disease. Within these groups the aim of this chapter is to illustrate, through case histories, how the bronchoscope may provide valuable additional information about these disease processes.

IMMUNOLOGICAL DISORDERS

Transbronchial biopsy is useful in diagnosing granulomatous disorders but less helpful when interstitial fibrosis is present. Lung lavage has been used in immunological disorders since 1977 after the work of Crystal in the United States. Lavage returns, obtained from patients with this variety of interstitial lung disease, can be divided into those where the percentage of lymphocytes is abnormally high (>10%) and those in which the neutrophil, not normally found, is more numerous (>5%), with or without increased eosinophil numbers.

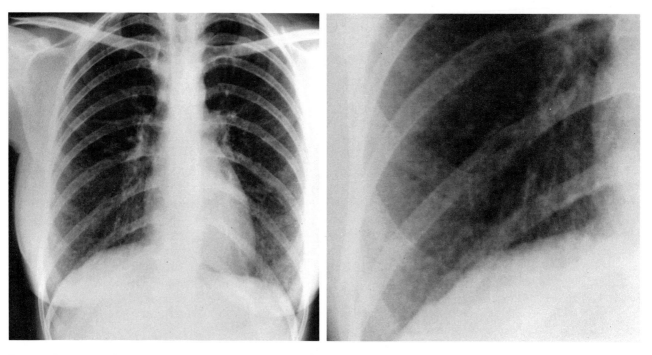

Fig. 4.2 Case study 1. Chest radiograph of a patient with sarcoidosis. Small nodular shadowing can be seen, particularly in the lower zones (see enlargement of right lower zone).

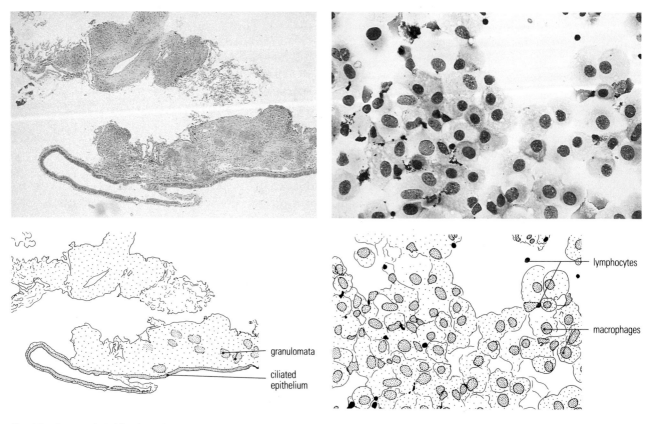

Fig. 4.3 Case study 1. Histology of a transbronchial biopsy in sarcoidosis. Multiple granulomata are present, predominantly in the subepithelial region. (H and E.).

Fig. 4.4 Case study 1. Lung lavage in sarcoidosis. A predominance of macrophages and some lymphocytes can be seen. (May-Grünwald-Giemsa.)

The former group includes patients with sarcoidosis and extrinsic allergic alveolitis, and in the latter, lymphocyte counts may exceed eighty per cent of the total cell yield. In sarcoidosis, the lymphocytes are predominantly helper cells, whereas in extrinsic allergic alveolitis more suppressor cells are obtained. Where fibrosis is present, neutrophil or eosinophil granulocytes are seen more frequently even in the granulomatous disorders which have progressed to fibrosis. Initial hopes that differential counts would provide diagnostic percentages have been disappointing, as there is considerable overlap between the diseases. This is especially true where sarcoidosis has progressed to severe fibrosis when neutrophils may be as plentiful as in fibrosing alveolitis.

Case Study 1
A thirty-year-old woman presented for a routine pre-employment chest radiograph. She had no symptoms, and on examination there were no abnormal findings. Her chest radiograph showed diffuse nodular infiltrates throughout both lungs (Fig. 4.2). A diagnosis of sarcoidosis was confirmed by transbronchial biopsy (Fig. 4.3) and lung lavage produced an excess of lymphocytes (Fig. 4.4).

Case Study 2
A thirty-year-old man presented with Raynaud's phenomenon, followed by small joint arthropathy and lymphadenopathy. He defaulted from follow-up but attended again two years later when the lymphadenopathy had progressed. Lymph node biopsy produced tissue containing non-caseating granulomata. He was treated with corticosteroids with improvement, but again defaulted from follow-up and presented once more, after another six years, with haemoptysis and a recurrence of arthropathy affecting predominantly the interphalangeal joints. On examination, he was found to have cervical lymphadenopathy, arthropathy of the proximal and distal interphalangeal joints of both hands but no other abnormalities. Chest radiography (Fig. 4.5) showed extensive fibrosis in addition to features compatible with a diagnosis of sarcoidosis, which was subsequently confirmed by transbronchial biopsy (Fig. 4.6). Lung lavage, however, produced a differential count which included nine per cent neutrophils and two per cent eosinophils, consistent with fibrotic development as suggested by the radiographic appearance.

Although an increase in lymphocytes is the predominant abnormality seen in lavage fluid from patients with sarcoidosis or extrinsic allergic alveolitis, it may also be found in fibrosing alveolitis. In this situation, it may occur either in lone cryptogenic fibrosing alveolitis (CFA) or in CFA associated with rheumatoid arthritis and other connective tissue disorders.

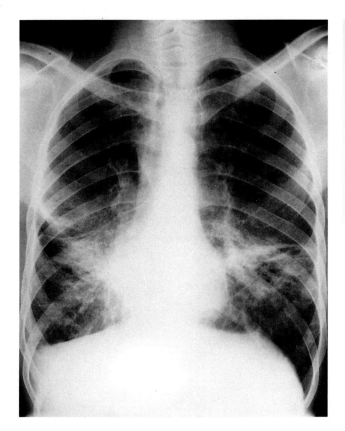

Fig. 4.5 Case study 2. Chest radiograph of a patient with sarcoidosis. Extensive mid-zone small nodular shadowing with linear shadowing in both mid- and lower zones can be seen.

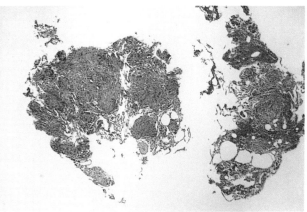

granulomata

Fig. 4.6 Case study 2. Transbronchial biopsy showing multiple epithelioid cell granulomata. (Haematoxylin and elastin van Gieson.)

Case Study 3

A sixty-five-year-old man complained of progressive breathlessness, without wheeze, of one year's duration. There were no other symptoms and, on examination, he was found to have no finger-clubbing; fine bilateral inspiratory crackles were heard at both lung bases. Chest radiography (Fig. 4.7) suggested a diagnosis of fibrosing alveolitis but lung lavage showed a high percentage of lymphocytes. A diagnosis of cryptogenic fibrosing alveolitis was confirmed at open lung biopsy.

Case Study 4

A sixty-four-year-old woman had suffered from rheumatoid arthritis for seven years. This had been treated with non-steroidal anti-inflammatory agents, but never with gold or penicillamine. She was referred to the chest clinic because of cough and breathlessness on exertion which had progressively worsened. On examination, there was no finger-clubbing but crackles were noted in the right and left anterior upper zones. Immunoglobulin analysis revealed a raised IgG but normal IgA and IgM; avian and aspergillus precipitins were absent. Angiotensin converting enzyme (ACE) levels were normal and a Mantoux test was negative. Her rheumatoid factor was positive to high titre. A chest radiograph (Fig. 4.8) showed predominantly upper lobe disease; transbronchial biopsy confirmed a diagnosis of interstitial fibrosis which was thought to be secondary to rheumatoid disease (Fig. 4.9). Lung lavage produced a high percentage of lymphocytes in addition to abnormal numbers of neutrophils and eosinophils.

There have been few large series of lung lavage in patients with fibrosing alveolitis related to connective tissue disorders, but studies of patients with systemic sclerosis have shown that lymphocytes do indeed feature

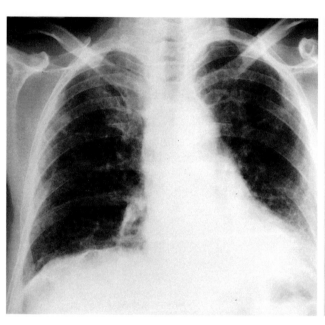
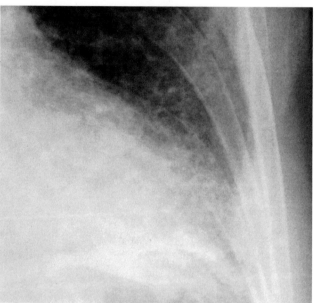

Fig. 4.7 Case study 3. Chest radiograph of a man with fibrosing alveolitis. Reticulonodular shadowing is seen throughout both lung fields but particularly at the left base (see enlargement).

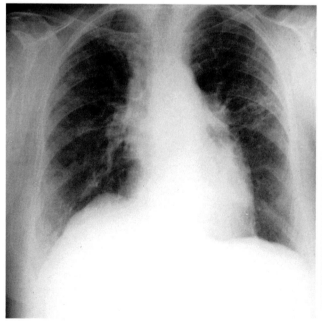
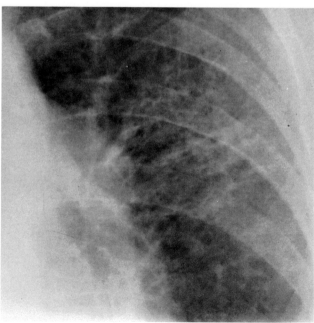

Fig. 4.8 Case study 4. Chest radiograph of a patient with rheumatoid lung. Bilateral changes consistent with bilateral upper zone fibrosis are seen. Note also the nodular shadowing in the left upper zone (see enlargement).

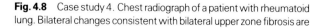

in the lavage returns from these patients. Lavage fluid immunoglobulin levels suggest that the lung may produce immunoglobulin locally and that the excess of lymphocytes may be a reflection of this enhanced local immunological activity. Using a panel of monoclonal antibodies directed against lymphocyte and macrophage subsets, Campbell and co-workers (1985) have demonstrated that true germinal centres are present in the lungs of patients with fibrosing alveolitis, supporting the hypothesis of local production. It is possible that raised serum levels of immunoglobulins and immune complexes may result from this local production.

In extrinsic allergic alveolitis, the lavage differential cell count is the most 'diagnostic test' in the immuno-

logical interstitial lung disorders. The lymphocytes are often the most numerous cells present in lavage returns and appear activated on routine light microscopy.

Case Study 5

A twenty-six-year-old man presented with a three-week history of increasing breathlessness on exertion. He had a cough productive of yellow sputum. His father had kept pigeons for sixteen years and the patient was involved in cleaning the pigeon loft and training the racing pigeons. On examination, there were no abnormal clinical signs. Chest radiography (Fig. 4.10) revealed a small nodular pattern which was predominant in the mid- and upper zones of both lungs. Avian precipitin tests were positive, showing four lines to pigeon antigen, three lines to budgerigar antigen and three lines to chicken antigen.

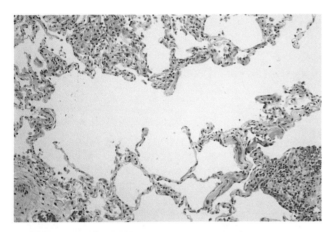

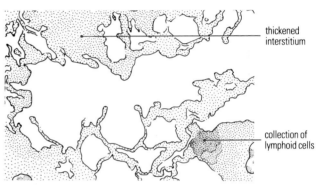

Fig. 4.9 Case study 4. Transbronchial biopsy showing alveolar wall thickening and an aggregation of lymphoid cells. (H and E.)

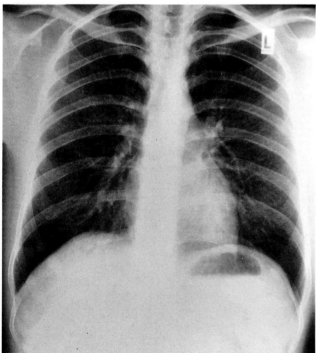

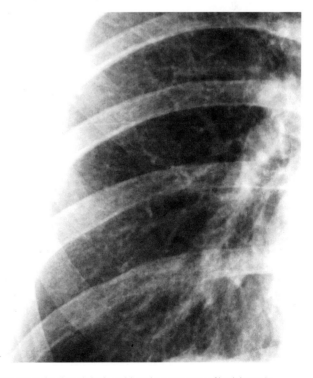

Fig. 4.10 Case study 5. Chest radiograph of a man with extrinsic allergic alveolitis due to exposure to pigeon antigens. A small nodular pattern predominantly in the mid- and upper zones of both lungs is shown (see enlargement of the right mid-zone).

Lung function tests confirmed a restrictive ventilatory defect with reduced gas transfer. Bronchoalveolar lavage was performed and a vast excess of lymphocytes was observed. Many of these appeared activated and, in some cytospin preparations, the lymphocytes seemed to cluster around the macrophages (Fig. 4.11).

Lung lavage can also be of great value in reinforcing a clinical suspicion of extrinsic allergic alveolitis.

Case Study 6

A seventy-three-year-old man had been under the care of another hospital for seven years. During the preceding three years he had been admitted on several occasions with breathlessness which was, at first, attributed to airflow obstruction and, latterly, to left ventricular failure. He was reviewed by the respiratory unit during one of these admissions and it was clear that his exercise tolerance had been reduced over recent years. At the time of his admission he was only able to walk twenty-five metres without breathlessness. He also had a history of angina which was relieved by nitrate therapy. On examination, he had evidence of finger-clubbing and also crackles at both lung bases and over both upper zones anteriorly. Chest radiography (Fig. 4.12)

showed a normal-sized heart with fine, reticulonodular shadowing in the upper zones; in addition, there were some linear shadows suggestive of fibrosis. Lung function tests revealed a restrictive ventilatory defect and avian precipitins were negative. In view of the pattern of radiological abnormality, lung lavage was performed and an excess of lymphocytes together with a modest elevation in neutrophil percentage was observed in the cell differential count (Fig. 4.13). The serum was re-examined and on this occasion budgerigar precipitins were discovered.

Lavage was extremely helpful in this patient's management as it was known that he had kept a budgerigar for eight years. However, as a widower, living alone, he had become extremely attached to this bird, and the initial finding of no precipitins encouraged his clinicians to advise him that he could keep the pet. Soon after discharge from hospital, his respiratory symptoms became worse and he required another hospital admission. It was on this occasion that lung lavage and the repeat avian precipitin test were performed. The lavage findings led to the final diagnosis being made and, although the neutrophils and radiological appearances suggested some fibrosis, it was hoped that removal of the source of antigen together with corticosteroid therapy would prevent further deterioration.

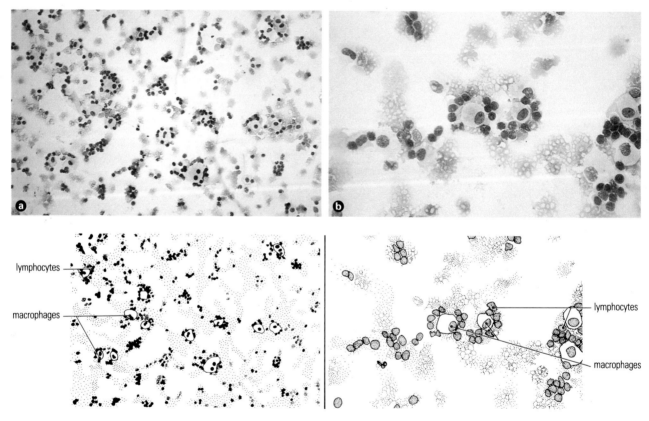

Fig. 4.11　Case study 5. Lung lavage in extrinsic allergic alveolitis. View (a) shows numerous lymphocytes, many of which are rosetting around macrophages. (b) is a high power view of rosetting lymphocytes. Cell differential count: macrophages 16%, lymphocytes 83%, neutrophils 1%. (May-Grünwald-Giemsa.)

Macrophages, seen after May-Grünwald Giemsa staining, appear to be a heterogeneous population and it was hoped that the identification of different antigens on the cell surface might enable a final distinction to be made between different diseases. This hope has not been sustained although macrophages from patients with immunological lung diseases do express a unique antigen profile when compared with controls. When cells from patients with sarcoidosis are compared with those obtained from patients with cryptogenic fibrosing alveolitis, however, no differences are seen. Nonetheless, the combination of cytochemical and monoclonal antibody techniques has confirmed the heterogeneity of the macrophage population (Fig. 4.14) but, more importantly, on tissue sections it is possible to identify subsets of macrophages and their *in situ* spatial relationship with other inflammatory cells (Fig. 4.15).

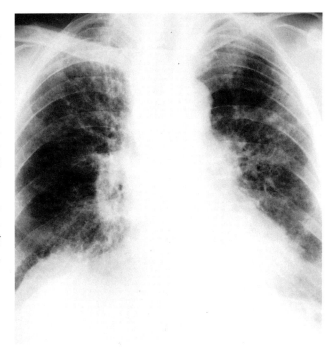

Fig. 4.12 Case study 6. Chest radiograph of a man with extrinsic allergic alveolitis due to exposure to budgerigar antigens. This shows small lung fields with linear shadows which are suggestive of fibrosis, best seen on the left upper zone. Nodular shadowing is present throughout both lung fields, but is most pronounced in the right and left upper zones.

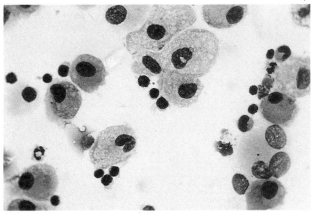

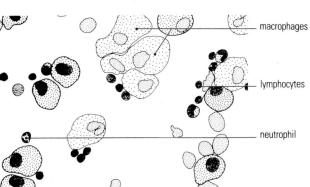

Fig. 4.13 Case study 6. Lung lavage showing an excess of neutrophils and lymphocytes. Cell differential count: macrophages 43%, lymphocytes 47%, neutrophils 8%, eosinophils 2%. (May-Grünwald-Giemsa.)

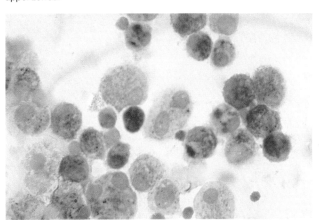

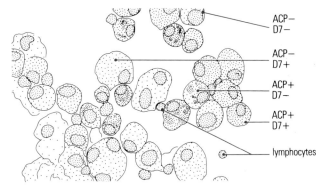

Fig. 4.14 Cytospin preparation from a patient with sarcoidosis. This is a combination preparation using an acid phosphatase reaction to detect intracytoplasmic acid phosphatase (ACP) and an immunoperoxidase method to identify the location of monoclonal antibody RFH D7 directed against cell surface antigen. The pink intracytoplasmic staining reveals acid phosphatase and the brown surface staining represents the site of RFH D7 antibody. It can be seen that there is a heterogeneous population of macrophages within this field and that the lymphocytes take up neither reagent.

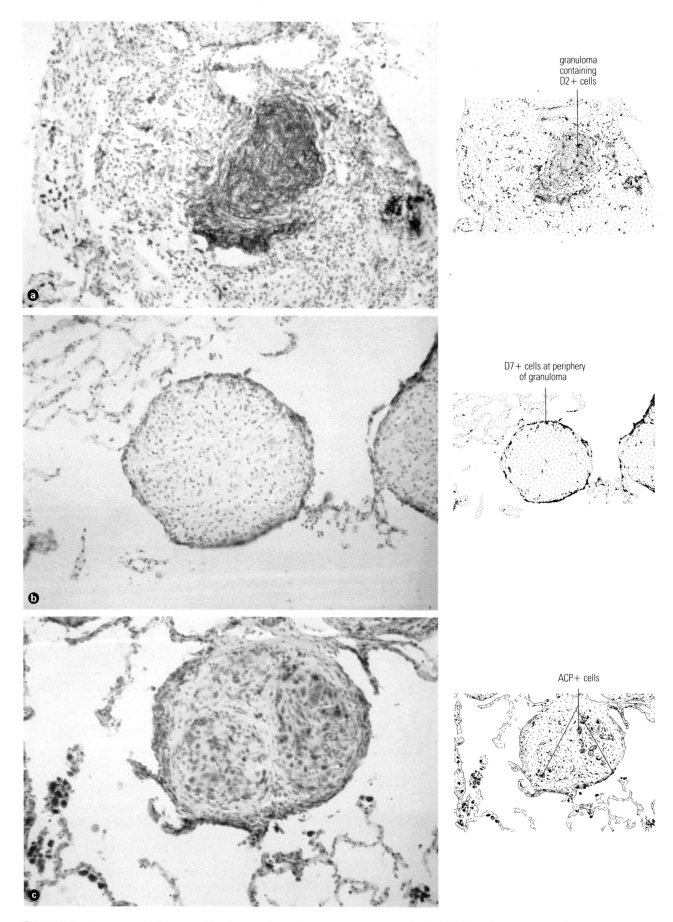

Fig. 4.15   Three transbronchial biopsy sections from a patient with sarcoidosis. View (a) has been treated with a monoclonal antibody RFH D2 directed against the epithelioid cells within the granuloma and visualized with an immunoperoxidase technique. View (b) shows a granuloma that has been treated with a different macrophage monoclonal antibody (RFH D7) which recognizes cells around the periphery of the granuloma. View (c) shows a granuloma that has been treated with reagents which identify acid phosphatase. This shows that the central (D2+) macrophages as well as the peripheral (D7+) macrophages all stain positively for acid phosphatase.

OCCUPATIONAL LUNG DISORDERS

Pollution of the environment has received increasing attention in recent years. Legislation is being introduced to control the working environment and similar controls on the general environment seem desirable. The major impact of environmental pollution is on the lungs, and sampling of tissue from transbronchial biopsy and lung secretions, and cellular constituents obtained by bronchoalveolar lavage can reveal exposure to various inorganic materials such as asbestos. Other particulate inclusions can be seen in macrophages and, using techniques such as scanning electron microscopy and electron probe analysis of cell inclusions, valuable information can be obtained about other interstitial lung diseases caused by environmental pollution at work. It seems likely that such analytical techniques will expand in specialist centres. At present, useful information can still be obtained using light microscopy, and an awareness of occupational causes for interstitial lung disorders must be maintained.

Case Study 7

A forty-four-year-old man presented with a two-year history of dyspnoea which was progressive, with no other chest symptoms. As a lagger, he had been exposed to chrysotile asbestos from the age of fifteen to thirty-three years. He had smoked until two years before assessment. On examination, he had finger-clubbing and bilateral basal crackles. Chest radiography (Fig. 4.16) revealed extensive nodular shadowing of both lungs, particularly at the bases where it had become confluent. There was marked pleural shadowing along the lateral chest walls, with calcification of the domes of both diaphragms. Lung function tests showed a restrictive ventilatory defect with reduced gas transfer. Lung lavage produced an excess of neutrophils and eosinophils and large numbers of asbestos bodies (Fig. 4.17).

Although the presence of asbestos bodies in lung lavage fluid does not allow a definite diagnosis of asbestosis, it is proof that the patient has been exposed to asbestos. The work of Bignon and his colleagues suggests that asbestos body and fibre counts may ultimately be used to assess the probability that pulmonary fibrosis results from asbestos exposure; however, the quantification of asbestos bodies is not yet accepted as a predictor of fibrosis.

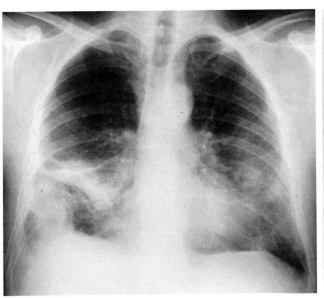

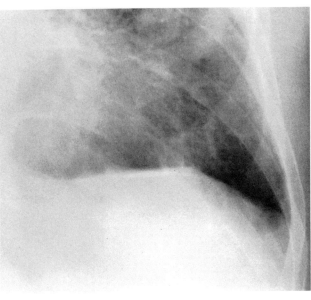

Fig. 4.16   Case study 7. Chest radiograph of a man with asbestosis. Nodular shadowing in the left lower zone and extensive pleural shadowing, particularly right-sided are seen. Diaphragmatic calcification, best seen on the left diaphragm, is also present (see enlargement). By courtesy of Dr. A. Newman-Taylor.

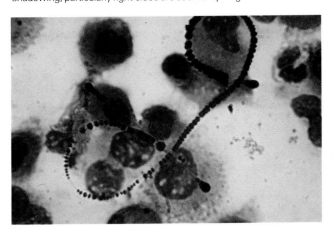

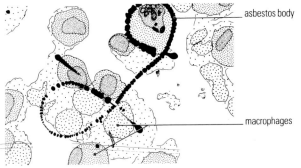

Fig. 4.17   Case study 7. Lung lavage sample showing a huge asbestos body wrapped around two macrophages. (May-Grünwald-Giemsa.) By courtesy of Dr. A. Newman-Taylor.

Case Study 8

A sixty-one-year-old man presented with a six-month history of progressive breathlessness on climbing stairs at home. As a cigarette smoker, he had experienced a cough productive of white sputum for many years, but there had been no recent change in its character. He had worked as a rubber moulder for twenty-one years; this entailed working on a rubber extrusion process using talc to prevent adhesion of the rubber to the extruder. Talc dust was present in the atmosphere and was almost certainly the source of his pneumoconiosis. There were no abnormal clinical signs. Chest radiography showed widespread nodular shadowing in both lungs consistent with pneumoconiosis (Fig. 4.18). Lung function tests showed mild airflow obstruction with a normal gas transfer factor. A transbronchial biopsy produced tissue containing heavy deposits of dust which were birefringent. The particles were thought to be talc (Fig. 4.19).

In this patient, transbronchial biopsy provided adequate results for a firm diagnosis to be made when his occupational history was taken into account. Other forms of inorganic dust exposure can be analysed using sophisticated microprobe analysis techniques; it is likely that these techniques, in combination with lung lavage or transbronchial biopsy, may allow more precise assessment of environmental exposure to be made.

Case Study 9

A thirty-five-year-old man developed cough and breathlessness after exertion seventeen months after starting work grinding pre-sintered hard-metal. Chest radiography was normal. Lung function tests showed a restrictive ventilatory pattern with reduced gas transfer factor. Bronchoalveolar lavage was performed; an excess of eosinophils were present and ten per cent of the macrophages had cytoplasmic particles

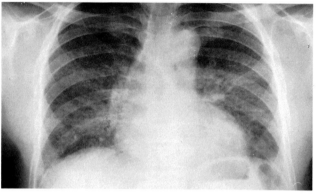

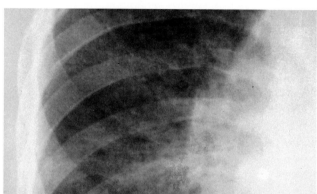

Fig. 4.18 Case study 8. Chest radiograph of a patient with talcosis. Widespread small nodular shadowing is present throughout both lung

fields (see enlargement of right mid-zone). By courtesy of Dr. A. Newman-Taylor.

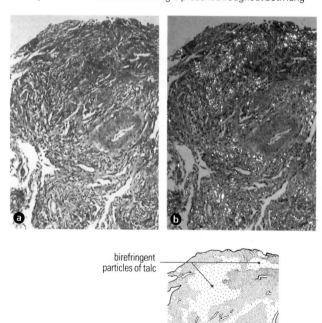

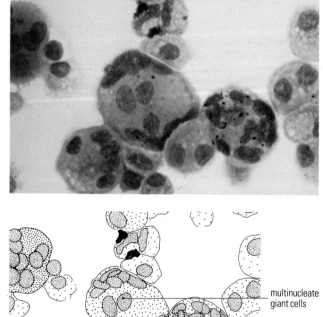

Fig. 4.19 Case study 8. Transbronchial biopsy in talcosis. Gross thickening of the interstitium is present with loss of the normal alveolar architecture. Section (b) is the same section as in (a) but viewed under cross-polarized light. Heavy deposits of dust are present and are seen to be birefringent. (H and E.) By courtesy of Dr. A. Newman-Taylor.

Fig. 4.20 Case study 9. Lung lavage preparation from a patient with hard-metal disease. Several multinucleate giant cells can be seen. (May-Grünwald-Giemsa.) By courtesy of the publishers, *Thorax*.

which were refractile but not birefringent. Six per cent of the macrophage population consisted of bizarre multinucleate giant cells with as many as twenty nuclei (Fig. 4.20). Electron microscopy showed that the giant cells had features of macrophages. Electron probe analysis of the lung lavage fluid revealed tungsten and iron. Hard-metal is an alloy of tungsten carbide and cobalt and is used to make drill tips, tool edges and armament components.

Interstitial lung disease due to work with hard-metal has been recognized since 1940. In the above case study, the relatively non-invasive technique of lung lavage provided material which allowed a precise diagnosis to be made.

Use of a basic lung lavage without further sophisticated analysis can also be helpful in other situations.

Case Study 10

A thirty-eight-year-old man noted a gradual onset of breathlessness for two years. For six years he had worked as a welder and did not wear any respiratory protection. On examination, there were no abnormal findings, but chest radiography (Fig. 4.21) revealed bilateral nodular shadowing throughout both lung fields. His lung function tests showed a mild, predominantly obstructive ventilatory defect. Bronchoalveolar lavage produced macrophages which were full of iron particles, and a transbronchial biopsy obtained tissue rich in iron (Fig. 4.22). The differential cell count was normal. A diagnosis of siderosis was made.

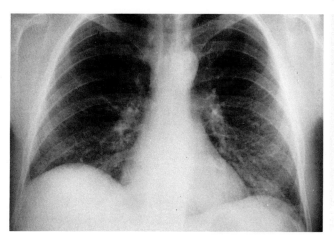
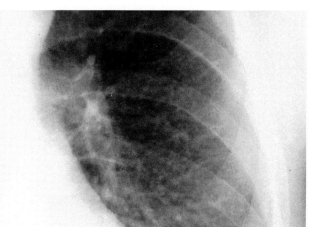

Fig. 4.21　Case study 10. Chest radiograph of a man with siderosis. Nodular shadowing can be seen in both lungs (see enlargement). By courtesy of Dr. A. Newman-Taylor.

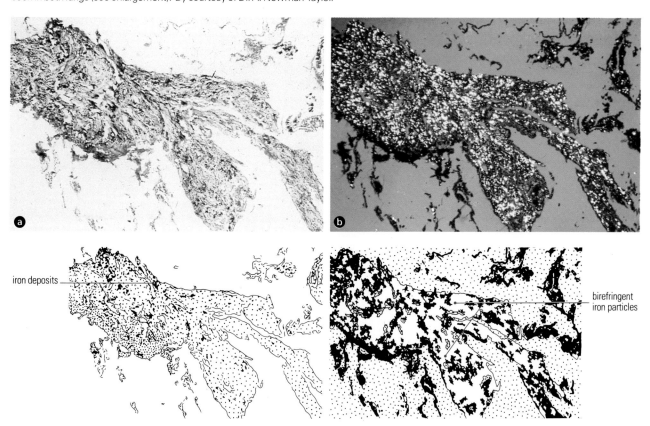

iron deposits

birefringent iron particles

Fig. 4.22　Case study 10. Transbronchial biopsy of a man with siderosis. View (a) shows alveolar thickening. View (b) is the same section as view (a) but taken with polarized light which demonstrates the iron particles that have impregnated the whole of the tissue. (H and E.) By courtesy of Dr. A. Newman-Taylor.

Many of the occupational lung disorders result in fibrosis and, for the reasons already outlined, transbronchial biopsy is often unrewarding. However, where the pathological processes are predominantly granulomatous, transbronchial biopsy can confirm a clinically suspected diagnosis.

Case Study 11

A fifty-two-year-old man presented for a routine pre-employment chest radiograph, and this was noted to be abnormal. He had a history of cough without breathlessness over the preceding five years. He gave a past history of bladder and left renal calculi. Three twenty-four hour urinary calcium levels were normal as was an intravenous pyelogram. The renal calculus was passed spontaneously and it consisted of calcium oxalate. Twenty years previously he worked in the chemical industry with powdered beryllium oxide for a duration of three years. He wore no respiratory protection. On examination, his fingers were clubbed, but there were no other abnormalities. Chest radiography showed widespread small nodular shadowing throughout both lung fields (Fig. 4.23). Measurements of serum and urinary calcium, ACE, aspergillus and avian precipitins and a Kveim test were undertaken; these were all negative or normal. Lung function tests showed a restrictive ventilatory defect with markedly reduced gas transfer factor. Transbronchial biopsy (Fig. 4.24) produced tissue containing multiple non-caseating granulomata; thus the diagnosis of chronic berylliosis was made. A beryllium lymphocyte transformation test performed on peripheral blood lymphocytes was positive.

In view of the diagnostic and medico-legal importance of identifying specific causes of occupational interstitial lung disease, it is likely that techniques such as lung lavage and transbronchial biopsy in association with sophisticated means of analysis of inorganic particles or fibres will be used more frequently.

INFECTION

Fibreoptic bronchoscopy is being used increasingly to help to establish the nature of pathogens in pulmonary infection. When blood cultures and sputum fail to identify a pathogen, bronchoscopy and selective brush sampling provides uncontaminated material from the lower respiratory tract. These techniques are even more useful for those patients who are immunosuppressed and who require rapid diagnosis of life-threatening infections. A combination of brushing (using a sealed, disposable brush), lavage and transbronchial biopsy may produce diagnostic samples.

Case Study 12

An eighteen-year-old woman developed a viral illness which failed to resolve. She remained tired with weight loss and increasing abdominal girth. On examination, she was clinically anaemic, had fundal haemorrhages and her spleen was enlarged 25cm below the costal margin. Haematological investigations resulted in a diagnosis of chronic granulocytic leukaemia, and she was treated with a

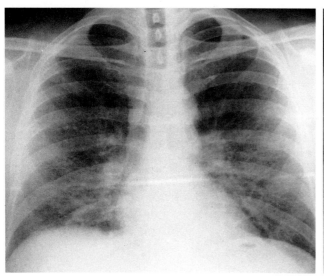

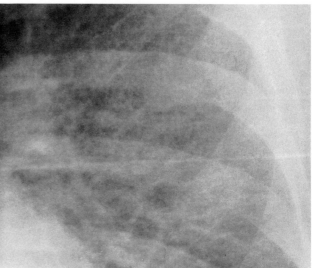

Fig. 4.23 Case study 11. Chest radiograph of a man with chronic berylliosis. Diffuse small nodular shadows are present throughout both lungs (see enlargement of left mid-zone).

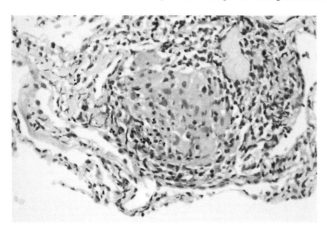

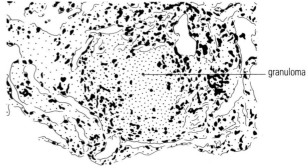

granuloma

Fig. 4.24 Case study 11. Transbronchial biopsy showing a non-caseating epithelioid cell granuloma. (H and E.)

combination of busulphan and leukopheresis with improvement. She remained in remission for two years, when an incipient blast cell transformation was treated with high dose melphalan followed by an autograft with peripheral blood, harvested during a chronic phase of her leukaemia. Within two weeks of her graft she developed a dry, non-productive cough with fever and a CMV titre showed a rise from less than 1 in 8 to 1 in 128. Lung function tests showed a restrictive ventilatory defect with low gas transfer. Chest radiography (Fig. 4.25) was consistent with an interstitial process and transbronchial biopsy was performed (Fig. 4.26). This showed viral inclusion particles consistent with a CMV infection.

In some situations, trauma must be avoided and bronchoalveolar lavage alone should be used; this is well tolerated by almost all patients despite their underlying disorder and risk of haemorrhage. Diagnostic yields from bronchoscopy are very variable in studies of immunosuppressed patients, and this reflects the heterogeneity of the underlying disorders and the stage of the illness at which bronchoscopy is performed. There have been no prospective studies designed to look at the relative yields from fibreoptic bronchoscopy, either performed immediately a lung problem is diagnosed or after a delay during which empirical therapy is tried. In several retrospective studies of bronchoscopy in the immunosuppressed patient, it was found that an organism is identified in approximately fifty per cent of patients depending on whether or not antimicrobial drugs have been given. It is possible that this percentage could be improved if bronchoscopy is employed earlier.

New advances in techniques used to process material obtained at bronchoscopy have increased the diagnostic yield. For example, a technique developed by Griffiths (1984) identifies intracellular CMV inclusions using monoclonal antibodies, making it now possible to diagnose CMV pneumonitis within twenty-seven hours instead of the mean 17.5 days using conventional tests based on cytopathic effect.

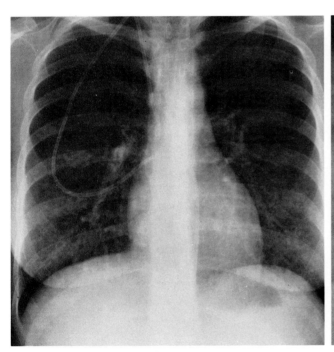
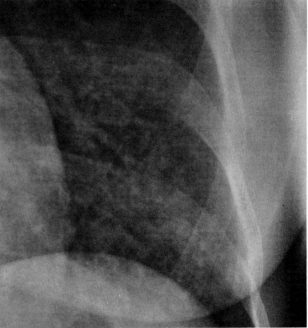

Fig. 4.25   Case study 12. Chest radiograph of a woman with CMV pneumonitis. Nodular shadowing in both lower zones can be seen (see enlargement).

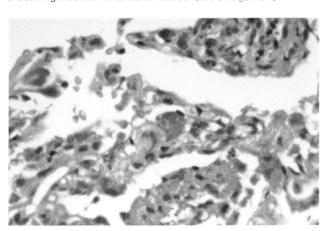
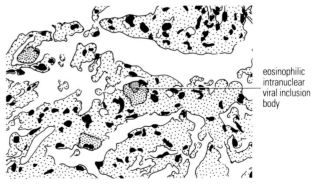

eosinophilic intranuclear viral inclusion body

Fig. 4.26   Case study 12. Transbronchial biopsy showing an eosinophilic inclusion body typical of CMV infection. (H and E.)

Case Study 13

A thirty-year-old woman presented with right-sided lymph-adenopathy. Mediastinal nodes were noted on chest radiography. Lymph node and bone marrow biopsy resulted in a diagnosis of T-cell leukaemia or lymphoma. She was treated with vincristine, adriamycin and prednisolone which produced remission. Five months later, she was given a preconditioning treatment with cyclophosphamide followed by total body irradiation and autologous bone marrow transplantation from her sister. CMV serology was positive before the transplant; one month after transplant, however, she developed a fever with inspiratory crackles at the left lung base. Chest radiography (Fig. 4.27) showed some patchy shadowing at the left lung base and

lung lavage was performed. Haemostasis problems prevented more invasive investigations. An initial CMV screen was negative but, after twenty-four hours of culture, CMV was identified by immunofluorescence (Fig. 4.28). She was treated with hyperimmune CMV serum and improved clinically and radiologically. She remains alive and well.

It is hoped that similar tests may be developed shortly which will enable other viruses to be identified as quickly. Other organisms, with the possible exception of fungi, are already speedily identified, and the importance of this in the immunosuppressed patient cannot be overemphasized.

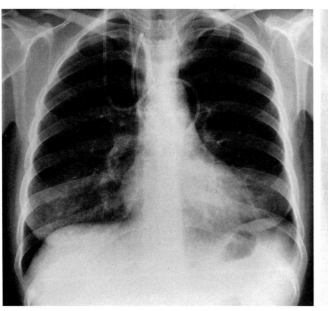

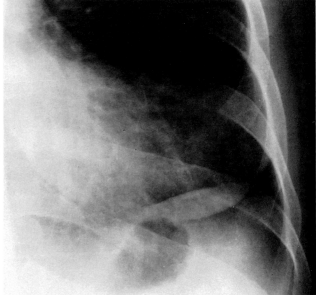

Fig. 4.27   Case study 13. Chest radiograph of a patient with CMV pneumonitis. Patchy shadowing of the left lung base is present (see enlargement).

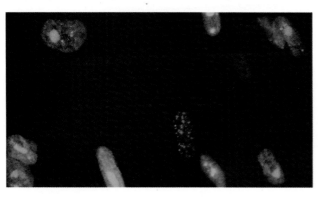

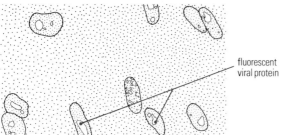

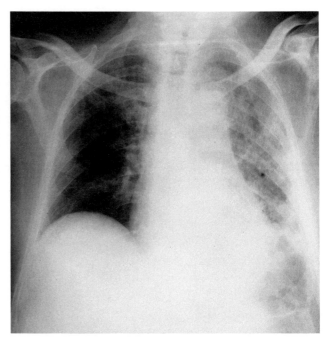

Fig. 4.28   Case study 13. Fibroblast culture showing fibroblasts within which immunofluorescent material is seen. Monoclonal antibodies (mouse immunogobulin) identify CMV protein within the fibroblasts. Fluorescein-labelled anti-mouse immunoglobulin reacts with this antibody, thereby locating the viral protein. By courtesy of Dr. P. Griffiths

Fig. 4.29   Case study 14. Chest radiograph of a man with Waldenström's macroglobulinaemia and pneumocystis infection. Extensive shadowing throughout the whole of the left lung and the right upper zone is shown.

Case Study 14

A seventy-four-year-old man was diagnosed as having Waldenström's macroglobulinaemia two years before presentation. No treatment was necessary for that two-year period but at the time of his second presentation hyperproteinaemia was found and raised blood viscosity was noted. He was treated with plasmaphaeresis and then chlorambucil. He developed the dermatological side-effects of chlorambucil; therefore this was replaced with cyclophosphamide. Seven months later, he noted increasing breathlessness and lethargy and, on examination, was found to have crackles in the right lower zone and throughout his left lung. Chest radiography showed extensive but asymmetrical shadowing (Fig. 4.29) and lung lavage was performed in the left upper zone. This revealed *Pneumocystis carinii* (Fig. 4.30).

The diagnosis of pneumocystis is becoming relatively easy to make by the use of bronchoalveolar lavage alone. It is necessary to perform lavage with volumes of at least 100ml in order to provide a good diagnostic yield. Using large volume lavage, a diagnosis will be made in over ninety per cent of patients with pneumocystis pneumoni-

tis. This relatively non-invasive technique has become more important with the increased incidence of acquired immune deficiency syndrome (AIDS), since it may be hazardous to perform a transbronchial biopsy in such patients because of the risk of pneumothorax.

Case Study 15

A thirty-three-year-old bisexual had lived in New York for three years. He had used marijuana and cocaine. Five months before presentation to our unit he had noted breathlessness and fever in New York, where he was treated with empirical antibiotics and discharged from hospital the following day. Five months later, he developed pleuritic chest pain with increasing breathlessness, weight loss and oral candidiasis. He was found to have lymphadenopathy in the left axilla and bilateral coarse crepitations at both lung bases. Investigations included a positive serum test for syphilis, and positive hepatitis B surface antigen. Peripheral blood T4 cells comprised six per cent of total T-lymphocytes and T8 cells, forty-two per cent of total T-lymphocytes. Blood gases on room air were pH 7.45, PaO_2 7.74kPa, $PaCO_2$ 2.4kPa. His chest radiograph (Fig. 4.31) was suggestive of a pneumonitis.

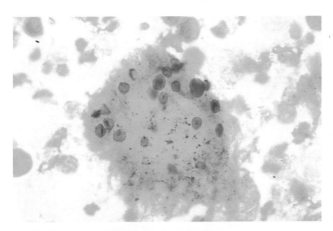

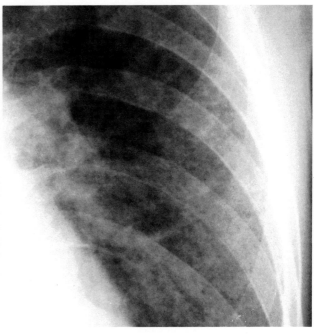

Fig. 4.30 Case study 14. Grocott stain of lung lavage material showing pneumocysts. By courtesy of Dr. C. Grubb.

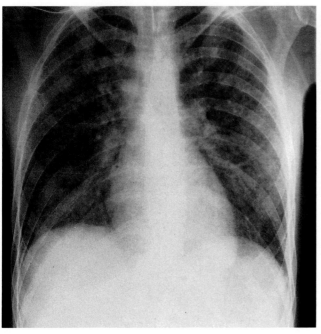

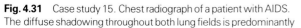

Fig. 4.31 Case study 15. Chest radiograph of a patient with AIDS. The diffuse shadowing throughout both lung fields is predominantly hazy and ground glass in appearance, but in areas has assumed a nodular character (see enlargement of left mid-zone).

Transbronchial biopsy was performed and this showed the presence of pneumocysts (Fig. 4.32). On the seventeenth day of his illness CMV was isolated in his urine and on day twenty, because of progressive respiratory failure, a repeat bronchoscopy was performed with bronchoalveolar lavage. This showed CMV inclusions in lung lavage cells identified by monoclonal antibody (Fig. 4.33).

Fibreoptic bronchoscopy is not without risk in patients who are severely ill and immunosuppressed. However, early bronchoscopy with platelet and fresh frozen plasma cover if necessary, will often allow a precise microbial diagnosis to be made. This enables specific therapy to be instituted and prevents the concurrent use of a number of potentially toxic drugs, in a 'blind' empirical treatment regimen.

DRUG REACTIONS

There are a large number of drugs which may affect the lungs. Some of these (e.g. methotrexate) are associated with a peripheral blood eosinophilia, but in general there are no specific features which identify a lung problem as being drug-induced. Although characteristic features may be seen in open lung biopsy samples, transbronchial biopsies generally provide too small a sample for a positive diagnosis. Medical literature contains a paucity of data on bronchoalveolar lavage in drug-induced pneumonitis. However, there have been reports of lung lavage cell differential counts on materials obtained from patients with amiodarone pneumonitis, and this reflects the value of a new technique being used in the diagnosis of patients with a recently described side-effect of a new drug.

Case Study 16

A fifty-six-year-old man with ischaemic heart disease presented with a two-week history of increasing breathlessness and pleuritic pain. One and a half years previously he had undergone coronary artery bypass grafting, and had commenced amiodarone treatment ten months later because of recurrent ventricular tachycardia. He was an ex-smoker and there was no significant exposure to asbestos or extrinsic allergens known to cause pulmonary disease. He had received a mean daily dose of amiodarone of 320mg for nine months and the only other drugs he was taking consisted of aspirin, frusemide and potassium supplements.

On examination, he was apyrexial and the only abnormal clinical signs were diffuse crackles over both lung fields. He showed no signs of finger-clubbing. Chest radiography revealed diffuse interstitial pulmonary shadowing with a normal cardiothoracic ratio (Fig. 4.34). Pulmonary function tests revealed a restrictive defect with a reduced single breath transfer factor. There was moderate hypoxaemia. A Swan-Ganz catheter revealed that the pulmonary capillary wedge pressure was not elevated.

A clinical diagnosis of amiodarone pulmonary toxicity was made and fibreoptic bronchoscopy was performed with transbronchial biopsy and bronchoalveolar lavage to attempt to confirm the diagnosis. The lavage return revealed a normal cell profile (with 95% macrophages, 4% lymphocytes and 1% neutrophils). Transbronchial lung biopsy (Fig. 4.35) showed numerous intra-alveolar 'foamy' macrophages and a variable inflammatory cell infiltrate of the alveolar walls. The histology was in keeping with a diagnosis of amiodarone pulmonary toxicity and the drug was stopped and prednisolone commenced at a dose of 60mg per day. The patient made a good initial response with an improvement in symptoms, chest radiographic appearances and pulmonary function.

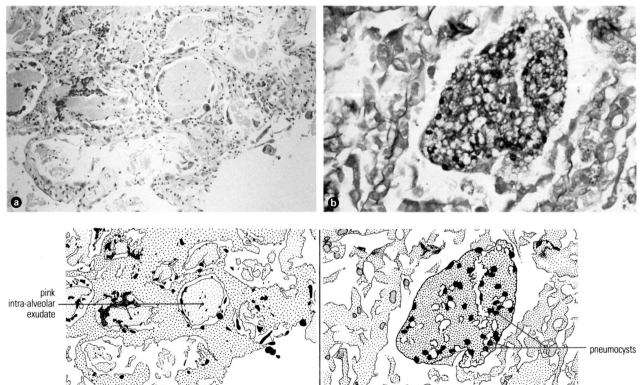

Fig. 4.32   Case study 15. Transbronchial biopsy showing alveoli filled with a pale homogeneous eosinophilic material. View (a) is an H and E section. Subsequent Grocott stain (b) confirmed that this material contained pneumocysts.

Data from other centres have revealed the presence of excess lymphocytes in the lavage fluid of patients with amiodarone pneumonitis. Lung lavage may also be of value when the patient has symptoms despite normal chest radiograph and lung function tests. An abnormal cell differential count can be helpful in deciding whether to proceed to more invasive investigations.

Case Study 17

A forty-nine-year-old man presented with a dry cough. Ten years previously, he had been treated with renal dialysis for renal failure due to polycystic disease and two years later had been given a cadaveric renal transplant. He had subsequently been immunosuppressed with prednisolone and azathioprine. Five years after receiving his transplant he presented with jaundice and a diagnosis of pre-sinusoidal portal hypertension was made. Two years later he developed a dry cough. At that time, he gave no history of breathlessness and there were no abnormal clinical signs. One year later his symptoms were unchanged, but this time he was noted to have bilateral fine basal crackles. There was no significant occupational or hobby history, and he took no other drugs regularly. Chest radiography and transbronchial biopsy were unhelpful but lung lavage showed an excess of lymphocytes. He was therefore subjected to open lung biopsy. This confirmed the presence of an interstitial pneumonitis with fibrosis and many type II pneumocytes, features consistent with a drug-induced problem. His immunosuppression was changed from azathioprine to cyclosporin A, and two months later bronchoalveolar lavage showed that his lymphocyte percentage had fallen from twenty-seven to seven per cent.

Azathioprine may rarely induce a pneumonitis and also pre-sinusoidal portal hypertension. This concurrent presentation with two rare side-effects of a single drug, taken together with the open lung biopsy data is good evidence that this drug was the cause of both problems. Management of this patient's problem was made easier by the identification of the pneumonitis using lung lavage.

It is unlikely that lung lavage will ever provide completely diagnostic material but, in situations where open lung biopsy is not possible, serial lavage before and after withdrawal of the drug may implicate the drug in the pathogenesis of the problem.

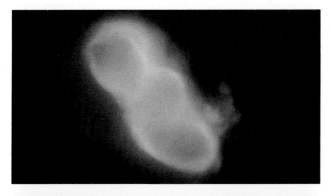

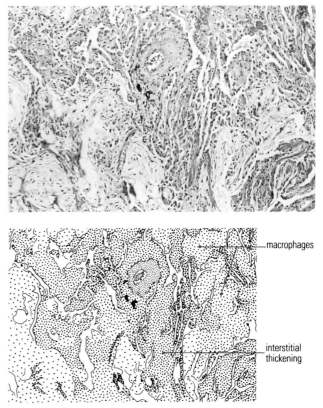

Fig. 4.33 Case study 15. Lung lavage preparation showing intracellular fluorescence. (CMV monoclonal antibody immunofluorescence technique.)

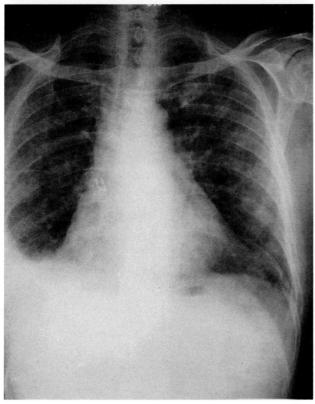

Fig. 4.34 Case study 16. Chest radiograph of a patient with amiodarone pneumonitis. Diffuse interstitial pulmonary shadowing is present.

Fig. 4.35 Case study 16. Transbronchial biopsy showing intra-alveolar macrophages and interstitial inflammation. (H and E.) By courtesy of Dr P. Corris.

RARER PROBLEMS

Several less common pulmonary disorders can be diagnosed by analysis of the lavage material. Disorders described in this section are unrelated pathogenically, but they are all examples of diagnoses which can be made by lavage.

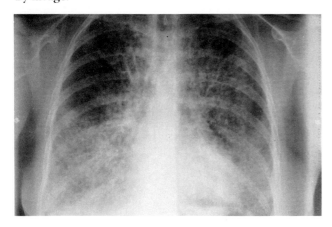

Fig. 4.36 Case study 18. Chest radiograph of a patient with alveolar proteinosis. Diffuse bilateral nodular changes can be seen.

Case Study 18

A forty-three-year-old woman presented with cough and breathlessness on exertion. Her chest radiograph revealed 'bilateral hilar lymphadenopathy' with some infiltrates but, despite a negative mediastinoscopy and transbronchial lung biopsy, a diagnosis of sarcoidosis was made. She was treated with prednisolone for nine months without improvement, and subsequently referred to a second hospital for assessment. At that time, her exercise tolerance was reduced to walking 200m on level ground. On examination she was cyanosed, but there were no other abnormal clinical signs. A second chest radiograph showed diffuse changes throughout both lung fields (Fig. 4.36). The PaO$_2$ was 4.4kPa breathing room air. She produced a small sample of sputum which showed characteristic lamellar bodies and, on the basis of this, a diagnosis of alveolar proteinosis was made. A therapeutic lung lavage was performed under general anaesthesia and this confirmed the diagnosis (Fig. 4.37). Further therapeutic lavages were performed with considerable improvement in her condition.

This patient's problems emphasized the need for a positive diagnosis. In her case, it was possible to obtain characteristic material from the sputum but often patients do not have sputum in this condition and a lung lavage will provide the material necessary for diagnosis.

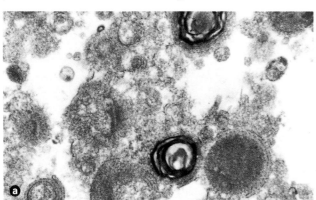

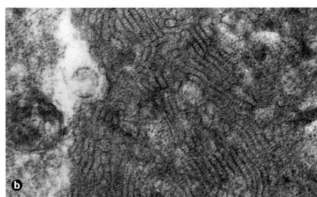

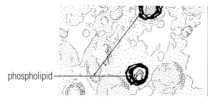

phospholipid

bilamellar phospholipid

Fig. 4.37 Case study 18. Electron microscopy of lung lavage material. The whorled appearance of the phospholipid inclusions within the cytoplasm of the macrophage (a) and the bilamellar appearance of the phospholipid in the lung lavage fluid (b) are shown.

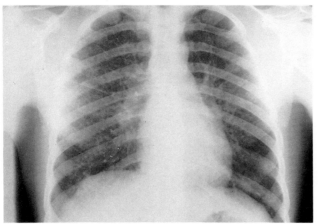

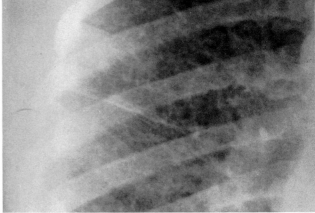

Fig. 4.38 Case study 19. Chest radiograph of a man with eosinophilic granuloma. Widespread reticulonodular changes can be seen in both lung fields (see enlargement of right mid-zone).

Case Study 19

A nineteen-year-old man gave a two-year history of a cough productive of a small amount of white sputum. In addition, he was breathless while riding his bicycle. Ten months before presentation he had had a right-sided pneumothorax which ultimately required thoracotomy to seal the air leak. At thoracotomy, his lung was abnormal and biopsy produced tissue consistent with eosinophilic granuloma. On examination, he had neither finger-clubbing nor focal signs in the chest. Lung function tests showed a mixed obstructive/restrictive ventilatory defect with reduced gas transfer but increased KCO. Chest radiography revealed widespread reticulonodular changes (Fig. 4.38). Broncho-alveolar lavage was performed and, although the differential cell count was normal, electron microscopy revealed typical inclusions in the macrophages (Fig. 4.39). These inclusions are pathognomonic for eosinophilic granuloma and if this patient had not undergone thoracotomy for therapeutic reasons, the diagnosis could have been made by lung lavage alone.

Case Study 20

A seventeen-year-old man had been well until three years before presentation when he was found to have an iron-deficiency anaemia with a haemoglobin of 7.5g/dl. Five months later he developed a cough with occasional haemoptysis. Although his anaemia had initially improved on iron supplements, it fell again three months later. Over the subsequent two years he had had repeated episodes of haemoptysis. On examination, there were no significant features in the chest. Chest radiography showed fine nodular shadowing throughout both lung fields (Fig. 4.40). Lung function tests were completely normal. Lung lavage produced numerous haemosiderin-laden macrophages (Fig. 4.41), thus a diagnosis of pulmonary haemosiderosis was made. There were no associated clinical features and particularly no evidence of Goodpasture's syndrome. This patient's history illustrates once more the value of lung lavage in one of the rarer interstitial lung problems.

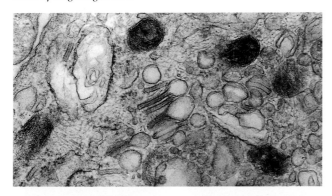

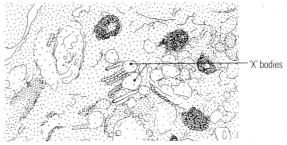

Fig. 4.39 Case study 19. Electron microscopy of cytoplasmic inclusions within a macrophage showing classical 'X' bodies.

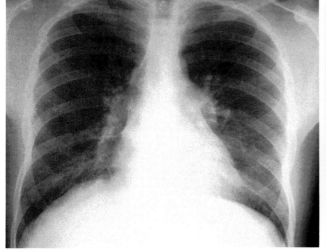

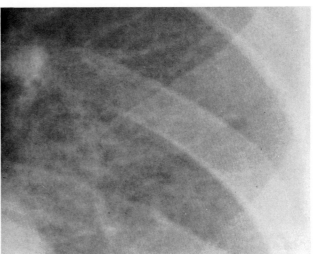

Fig. 4.40 Case study 20. Chest radiograph of a young man with haemosiderosis. Very fine nodular shadowing is present throughout both lung fields (see enlargement of left lung). By courtesy of Professor M. Turner-Warwick.

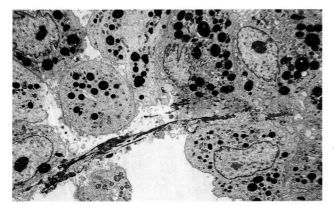

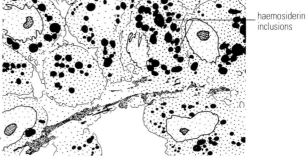

Fig. 4.41 Case study 20. Electron microscopy of several macrophages showing dense cytoplasmic inclusions of haemosiderin.

Another area in which lung lavage is often the only diagnostic technique available is the intensive care unit. Patients with severe multi-organ failure cannot be considered for invasive investigative procedures, but often develop pulmonary problems whilst there. Lung lavage can often provide a diagnosis of their pulmonary abnormalities.

Case Study 21

A twenty-seven-year-old man gave a five-week history of pain in the left humerus. Haematological assessment confirmed a diagnosis of acute lymphoblastic leukaemia and this was treated with combination chemotherapy. Two months after treatment, he became febrile with a rapid development of bilateral pulmonary shadowing (Fig. 4.42) and required ventilatory support for respiratory failure. *Streptococcus viridans* was subsequently isolated from his blood. Chest radiography and blood gas analysis suggested a diagnosis of adult respiratory distress syndrome in the setting of an immunosuppressed patient with septicaemia. Pulmonary capillary wedge pressure was normal. Bronchoalveolar lavage was performed and pink frothy material was returned. The cellular fraction of the lavage fluid contained an excess of neutrophils (Fig. 4.43) and was sterile. After treatment with high-dose corticosteroids and continued ventilatory support, he improved. At present the patient is alive and well.

Malignant processes may appear as a nodular or lymphatic pattern on chest radiography. Malignancy is considered further in chapter 3.

Patients with interstitial lung disease may present a difficult diagnostic problem. A positive diagnosis should be made in all of these patients, as an alternative to commencing an arbitrary treatment regimen (e.g. with corticosteroids) with its attendant side-effects. The relatively new techniques which can be applied using the fibreoptic bronchoscope have saved many patients from open lung biopsy. Where these techniques have failed to provide a diagnosis, open lung biopsy is usually necessary; however, with more advanced technology to process material obtained with the bronchoscope, open lung biopsy may be required less frequently in the future.

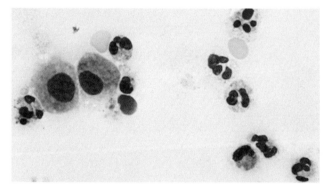

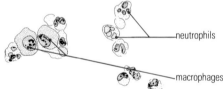

neutrophils

macrophages

Fig. 4.42   Case study 21. Chest radiograph of a man with adult respiratory distress syndrome. Patchy shadowing is present throughout both lung fields.

Fig. 4.43   Case study 21. Lung lavage from a patient with adult respiratory distress syndrome showing a gross excess of neutrophils. Cell differential count: macrophages 10%, neutrophils 90%. (May-Grünwald-Giemsa.)

REFERENCES

Campbell DA, Poulter LW, du Bois RM (1985). Immunocompetent cells in bronchoalveolar lavage reflect the cell populations in transbronchial biopsies in pulmonary sarcoidosis. *American Review of Respiratory Disease,* **132,** 1300-1306.

Crystal RG, Bitterman PB, Rennard SI, Hance AJ, Keogh BA (1984). Interstitial lung diseases of unknown cause: disorders characterized by chronic inflammation of the lower respiratory tract. Parts one and two. *New England Journal of Medicine,* **310,** 154-166 and 235-244.

Daniele RP, Elias JA, Epstein PE, Rossman MD (1985). Bronchoalveolar lavage: role in the pathogenesis, diagnosis and management of interstitial lung disease. *Annals of Internal Medicine,* **102,** 93-108.

Davison AG, Haslam PL, Corrin B, Coutts II, Dewar A, Riding WD, Studdy PR, Newman-Taylor AJ (1983). Interstitial lung disease and asthma in hard-metal

workers: bronchoalveolar lavage, ultrastructural and analytical findings and results of bronchial provocation tests. *Thorax,* **38,** 119-128.

Griffiths PD, Panjwani DD, Stirk PR, Ball MG, Ganczakowski M, Blacklock HA, Prentice HG (1984). Rapid diagnosis of cytomegalovirus infection in immunocompromised patients by detection of early antigen fluorescent foci. *Lancet,* **2,** 1242-1245.

Haslam PL, Turton CWG, Heard B, Lukoszek A, Collins JV, Salsbury AJ, Turner-Warwick M (1980). Bronchoalveolar lavage in pulmonary fibrosis: comparison of cells obtained with lung biopsy and clinical features. *Thorax,* **35,** 9-18.

Johnson N, Haslam PL, Dewar A, Newman-Taylor AJ, Turner-Warwick M (In press). Identification of inorganic dust particles in bronchoalveolar lavage macrophages by X-ray microanalysis. *Archives of Environmental Health.*

Mitchell DM, Mitchell DN, Collins JV, Emerson CJ (1980). Transbronchial lung biopsy through fibreoptic bronchoscope in diagnosis of sarcoidosis. *British Medical Journal,* **280,** 679-681.

Stover DE, Zaman MB, Hajdu SI, Lange M, Gold J, Armstrong D (1984). Bronchoalveolar lavage in the diagnosis of diffuse pulmonary infiltrates in the immuno-suppressed host. *Annals of Internal Medicine,* **101,** 1-7.

Turner-Warwick ME, Haslam PL (1986). Clinical applications of bronchoalveolar lavage: an interim view. *British Journal of Diseases of the Chest,* **80,** 105-121.

Wardman AG, Cooke NJ (1984). Pulmonary infiltrates in adult acute leukaemia: empirical treatment or lung biopsy? *Thorax,* **39,** 647-650.

Weinberger SE, Kelman JA, Elson NA, Young RC, Reynolds HY, Fulmer JD, Crystal RG (1978). Bronchoalveolar lavage in interstitial lung disease. *Annals of Internal Medicine,* **89,** 459-466.

5. Physiological Aspects of Bronchoscopy

M. D. L. Morgan MD MRCP

PHYSIOLOGICAL CONSEQUENCES OF BRONCHOSCOPY

In the last decade, the introduction of the fibreoptic instrument has revolutionized bronchoscopy. The procedure is now convenient for the physician, relatively acceptable to the patient and usually avoids the necessity for operating theatre time and anaesthetic assistance; it has proved to be extremely safe and free from major side-effects. A detailed discussion of the potential dangers of physiological disturbance by bronchoscopy would therefore appear to be academic. However, as confidence with the technique increases, its use is being extended to patients with poorer lung function and also prolonged and more complicated procedures are being employed. During rigid bronchoscopy, the airway is protected and oxygenation is maintained by the anaesthetist, but the fibreoptic bronchoscopist alone is responsible for the safety of the patient. It is therefore important for the bronchoscopist to understand the physiological consequences of the procedure and to be aware of the action necessary to correct any adverse effects.

The disturbances of physiology which accompany fibreoptic bronchoscopy are the consequence of many factors, some of which are present even before the bronchoscope is introduced into the airways. Fortunately many of the abnormalities of pulmonary function that can be detected are of no practical consequence but are discussed here for completeness. However, hypoxaemia and cardiovascular disturbance are potentially serious complications which are largely preventable and they will therefore be considered in detail.

Hypoxaemia

Bronchoscopy interferes with gas exchange and often produces a fall in arterial PO_2 to below 10.7kPa (80mmHg). This occurs in subjects with normal lungs as well as in patients with pre-existing hypoxaemia and is not usually accompanied by a rise in $PaCO_2$. The magnitude of the fall in PaO_2 varies with the individual. The average drop is about 2.7kPa (20mmHg), but falls of 5.3kPa (40mmHg) have been recorded. Often the largest falls in PaO_2 occur in those patients with the lowest starting values.

Hypoxaemia is not constant throughout the bronchoscopy. The lowest values of oxygen tension occur at the beginning of the procedure, when the instrument is in the pharynx and passes through the vocal cords, and again at the end when the fibrescope is withdrawn (Fig. 5.1). Severe hypoxaemia is also associated with lavage when the degree of disturbance roughly correlates with the volume of fluid which is inserted. Finally, the fall in oxygen tension does not revert to normal immediately, and significant hypoxaemia may continue for up to four hours following bronchoscopy, or even longer following lavage.

Mechanisms of hypoxaemia during bronchoscopy

The mechanisms which contribute to hypoxaemia are summarized in Figure 5.2. In the spontaneously breathing patient, the fall in PaO_2 is not accompanied by hypercapnia, although this may occur in unusual cases either when the fibrescope induces generalized bronchospasm or when there is severe pre-existing airway obstruction. In most circumstances, alveolar ventilation is maintained and the observed fall in PaO_2 cannot be attributed to the physical obstruction of the airway by the fibrescope. It is therefore likely that the abnormality of gas exchange is due to an alteration in ventilation/perfusion ($\dot{V}/\dot{Q}$) relationships within the lung.

These abnormalities are present and can be detected even before the fibrescope is introduced. The pre-medication drug and also assumption of the supine posture are associated with small falls in PaO_2. Once the fibrescope has been introduced, evidence from $\dot{V}/\dot{Q}$ radioisotope scans has suggested that low $\dot{V}/\dot{Q}$ areas are induced in the lung either by vagal reflexes secondary to tracheal stimulation or by lavage fluid. Saturation of a region of lung by saline lavage can induce a low $\dot{V}/\dot{Q}$ relationship simply by mechanical obstruction of the alveoli; this effect is proportional to the quantity of fluid that is introduced. Evidence for an additional reflex mechanism is supported by worsening of hypoxaemia induced by cold rather than warm lavage fluid and also by the development of $\dot{V}/\dot{Q}$ abnormalities in the contralateral lung during lavage.

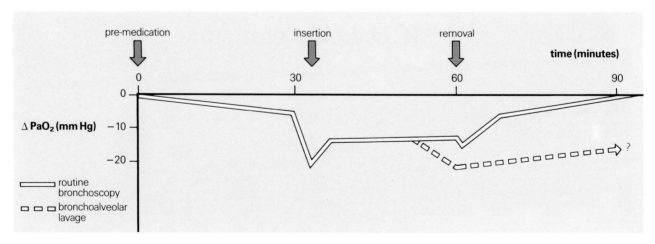

Fig. 5.1 Hypoxaemia in relation to bronchoscopy. Variations in PaO₂ during the various stages of bronchoscopic investigation.

Bronchoalveolar lavage may be associated with more profound and persistent hypoxaemia.

The effect on gas exchange during bronchoscopy is different when the patient is being ventilated by intermittent positive-pressure ventilation (IPPV). In this circumstance, the bronchoscopy is usually performed as a therapeutic measure to remove secretions, but it may also be done either for diagnostic purposes or when the patient requires anaesthesia. The calibre of the airway is already reduced by the endotracheal tube and further ventilatory obstruction results in a rise in $PaCO_2$. Also, because of mechanical ventilation, the patient is unable to regulate his or her own tidal volume. Serious interference with ventilation can be introduced by suction steal of the tidal volume. The rate of suction through the channel of the fibreoptic bronchoscope can reduce a 700ml tidal volume by forty per cent at $12.5cmH_2O$ of vacuum and seventy-five per cent at $21cmH_2O$ of vacuum. This hazard can be avoided by increasing the ventilation in anticipation of bronchoscopy.

Hypoxaemia induced by disturbance of the $\dot{V}/\dot{Q}$ relationship is less of a problem in the ventilated patient and can be covered by increasing the inspired oxygen concentration. In most cases, however, the improvement in oxygenation produced by the restoration of $\dot{V}/\dot{Q}$ by therapeutic lavage outweighs the risk of bronchoscoping the hypoxaemic, ventilated patient.

Monitoring oxygenation and delivery of supplementary oxygen

The delivery of oxygen to respiring tissues does not depend solely upon the gas exchanging properties of the lungs. The quantity of haemoglobin, the cardiac output and the vulnerability of potentially ischaemic regions of the brain or myocardium should all be considered before the patient has the bronchoscopy. If these conditions are known and stable, the consequences of the fall in arterial PO_2 induced by bronchoscopy will depend upon the starting level of oxygenation. Because of the oxygen carrying properties of haemoglobin, a fall in PaO_2 of 2.7kPa (20mmHg) will make no difference to delivery of oxygen to the tissues in the normal subject. However, once the resting level of PaO_2 has been reduced to 6.7kPa (50mmHg) by pulmonary disease, further small reductions in PaO_2 will result in dramatic falls in oxygen saturation (SaO_2) and oxygen delivery. A patient with normal resting blood gases will not therefore sustain any ill-effects even though they become mildly hypoxaemic ($PaO_2 < 10.7kPa - 80mmHg$). By contrast, a patient with a resting PaO_2 of 9.3kPa (70mmHg) is in danger of marked desaturation by a similar fall of PaO_2.

The effect of bronchoscopy on gas exchange can be observed either by intermittent measurement of arterial PO_2 or by the continuous measurement of oxygen saturation. There are obvious disadvantages to the use of intermittent collection of arterial blood gas tensions since they require vessel cannulation and cannot be analysed immediately. However, the measurement of ear-lobe saturation is rapid, unobtrusive and supplies the relevant information. This is therefore the method of choice for monitoring the oxygenation of patients when necessary.

Since the major mechanism of hypoxia is a disturbance in the $\dot{V}/\dot{Q}$ relationship, it should be possible to correct the abnormality in most cases by increasing the inspired concentration of oxygen. Several studies have shown this to be true and that most methods of oxygen delivery are equally satisfactory. Oxygen can be supplied via a face mask or by a nasal cannula to the unoccupied nostril. It is cumbersome to bronchoscope a patient through a face mask, although modification of the mask with a rubber diaphragm has been reported to produce satisfactory results. The single nasal cannula, however (Fig. 5.3), supplies sufficient oxygen and is unobtrusive. Methods of oxygen delivery employing direct flow down the suction channel of the fibrescope have not proved to be popular and in one instance was implicated in the production of a pneumothorax.

As a recommendation, patients undergoing bronchoscopy should have their arterial blood gases measured before the procedure if there is any significant impairment of respiratory function (forced expiratory volume – FEV_1 <1 litre). If the PaO_2 is below 9.3kPa (70mmHg), they should receive supplementary oxygen by nasal cannula and this should be continued for at least four hours after a prolonged procedure or large volume lavage. Monitoring of oxygenation by ear oximeter is not usually necessary, but should be reserved for those patients with either severe impairment of lung function or excessive vulnerability to falls in PaO_2.

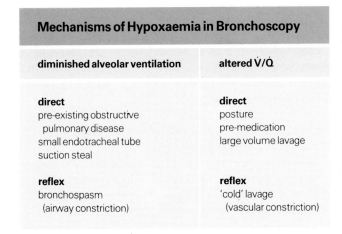

Mechanisms of Hypoxaemia in Bronchoscopy	
diminished alveolar ventilation	altered $\dot{V}/\dot{Q}$
direct	**direct**
pre-existing obstructive pulmonary disease small endotracheal tube suction steal	posture pre-medication large volume lavage
reflex	**reflex**
bronchospasm (airway constriction)	'cold' lavage (vascular constriction)

Fig. 5.2 Mechanisms of hypoxaemia during bronchoscopy. A summary of the contributory factors which result in hypoxaemia during bronchoscopic investigation.

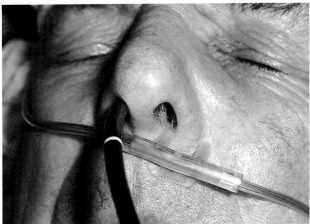

Fig. 5.3 Fibrescope and nasal cannula. The unobtrusive cannula supplies sufficient oxygen. The fibrescope is passed down the unoccupied right nostril.

Bronchoscopy and the Circulation

The safety record of fibreoptic bronchoscopy is good. Several large surveys of the complications of bronchoscopy indicate that the associated mortality of the procedure is about 0.02% (1:5,000). This figure is set against the background population of elderly patients who have higher than average incidences of ischaemic heart disease and chronic pulmonary disease. However, of fifteen deaths that occurred in published series, ten were attributed to a cardiovascular cause which may have been possible to identify and correct before cardiac arrest occurred.

There are two components to the cardiovascular disturbance which occurs during bronchoscopy: (i) the normal physiological response to the instrumentation and (ii) the haemodynamic consequence of pathological events which are either initiated by or coincident with the bronchoscopy.

The normal physiological response to bronchoscopy has not been studied in detail but some information is available. The passage of the fibrescope produces a fifty per cent increase in cardiac output which is associated with an increase in heart rate and a thirty per cent rise in systemic blood pressure. The pulmonary artery pressure also increases markedly. These events may all be related to catecholamine release arising from tracheal stimulation, hypoxaemia and apprehension and the effects can be attenuated by pre-medication with beta blocking drugs. These drugs are not, however, recommended because of their effect on airway obstruction.

Pathological haemodynamic disturbance occurs as a consequence of either myocardial infarction or the induction of serious cardiac arrhythmias. The sinus tachycardia induced by sympathetic activity will increase myocardial oxygen consumption. When this is associated with hypoxaemia in patients with coronary artery disease, there is a risk of myocardial infarction or arrhythmia. ST segment depression has been observed in patients with ischaemic heart disease coinciding with the hypoxic periods of the procedure. It is therefore prudent to give oxygen to the vulnerable patient and even to monitor the ECG during the high risk procedure.

Transient cardiac dysrhythmias are very common in patients undergoing bronchoscopy, even in those without evident ischaemic heart disease. The incidence of serious dysrhythmia varies from eleven to forty per cent of patients and cannot be predicted, in general, from the pre-operative assessment. Both atrial and ventricular dysrhythmias are observed and are usually well tolerated. Atrial dysrhythmias (ectopics and fibrillation) are neither related to the stage of the procedure nor to hypoxaemia. By contrast, ventricular ectopic activity occurs most frequently during passage through the vocal cords and persists after the end of the bronchoscopy in association with hypoxaemia. Reported theories of the cause of these dysrhythmias include: general anaesthesia if it is used, hypoxaemia, catecholamine release or excessive absorption of local anaesthetic. The latter theory is unlikely as absorbed levels are low and the effect of anaesthetic is generally antidysrhythmic. In spite of the frequency of cardiac dysrhythmias, none of the recorded series have reported haemodynamic disturbance of any consequence and no procedure has been abandoned because of dysrhythmia.

Lung Mechanics and Bronchoscopy

Insertion of the fibrescope into the bronchial tree will inevitably obstruct the airway. The importance of this obstruction will depend upon the relative sizes of the fibrescope and the airway. A small fibrescope in the adult trachea will have little effect but a large instrument passing through a small endotracheal tube will cause serious obstruction. The consequences of mechanical obstruction of the airway by the fibrescope will therefore be different with the transnasal approach than in the intubated or ventilated patient.

During a transnasal bronchoscopy, the standard fibrescope (external diameter usually 5.7mm) will reduce the size of the adult trachea by eleven per cent (Fig. 5.4). This has little effect on airway resistance during tidal breathing, only increasing airway pressures to approximately $3cmH_2O$ on expiration and $-5cmH_2O$ on inspiration. This small degree of obstruction is, however, sufficient to raise the functional residual capacity (FRC) by seventeen per cent during bronchoscopy. The relief of obstruction and reduction of FRC after the procedure may be an additional factor in postoperative hypoxaemia.

Endotracheal intubation, which is sometimes recommended in routine bronchoscopy, commonly with a tube of 8mm internal diameter, will on its own reduce the tracheal area by about fifty per cent and increase FRC. Further reduction of the original airway calibre by twenty-five per cent occurs when the fibrescope is inserted (Fig. 5.5). This degree of obstruction doubles the airway pressures during tidal breathing and increases the work of breathing. In addition, the obstruction to expiratory flow heightens the risk of barotrauma during coughing in the conscious patient.

It is now common practice to bronchoscope seriously ill patients who are being ventilated with IPPV. In these circumstances, the patient is already intubated and the ventilator can easily overcome the additional inspiratory resistance of the fibrescope. However, expiration is passive and the effect of the obstruction will be to preserve positive end-expiratory pressure (PEEP). The values of PEEP achieved by insertion of the fibrescope alone can reach $20cmH_2O$ or even higher if the endotracheal tube has an internal diameter less than 8mm. This will in some way offset the hypoxaemia produced by the bronchoscopy, but if PEEP is already being applied it must be discontinued before bronchoscopy to prevent lung rupture. When the tidal volume is fixed by the ventilator, the bronchoscopy suction may result in diminished alveolar ventilation. These losses can be considerable and the ventilator setting and the inspired oxygen concentration should therefore be increased before the procedure.

PHYSIOLOGICAL MEASUREMENT WITH THE FIBREOPTIC BRONCHOSCOPE

Unlike some other organs, there are measurable differences of function between different regions of the lung. This is because the alveolar/capillary unit structure makes the lung, as a whole, subject to the influences of gravity and posture. As a result there are regional differences in the distribution of ventilation and perfusion and in the arrangement of mechanical stresses. Measurements of regional lung function are of interest from two aspects: first, knowledge of the topographical differences in health and disease increases our understanding of the physiology of the lung; secondly, regional lung function measurements can influence clinical management, especially in the pre-operative assessment of patients for surgery. When a part of a lung has to be removed, the usual whole lung function tests are of limited value because they can neither provide local information about the part that has to be removed nor make any predictions about the postoperative behaviour of the remaining lung. There are several methods which are available to measure the regional differences in lung function; they include conventional radiography, radioisotope studies and, more recently, computed tomography (CT).

Access to the lung can also be obtained via the bronchial tree and functional measurements made through bronchoscopic sampling have the advantage of anatomical accuracy which is relevant to surgery. The history of bronchial cannulation for physiological measurement is well established. It is over a century since divided lung experiments were performed on animals and the first estimation of regional gas concentrations was made in man by Loewy and Schrotter in 1905. Since then, several varieties of double or triple-lumen catheters have been developed for bronchospirometry.

More recently the marriage of the respiratory mass spectrometer, developed by Fowler in the 1950's, to the rigid bronchoscope allowed West and Hugh-Jones to make accurate and continuous measurements of local gas concentration. The type of measurements which can now be made with the rigid bronchoscope include gas flow and analysis of expired gas content, pressure and temperature. Many of the techniques that were pioneered with the rigid instrument have now been adapted for use with the fibreoptic bronchoscope. This extends the anatomical range of measurement as well as reducing the unnatural interference of the measuring technique. The remainder of this chapter will consider the value of the fibreoptic bronchoscope in making clinically useful estimates of regional lung function and also the nature of its role in physiological reseach.

Bronchoscopic Lung Function Tests
Two techniques that can be useful in the pre-operative assessment of patients have been adapted for use with the fibreoptic bronchoscope: one is bronchospirometry and the other is the analysis of expired gas concentrations. Between them they can provide regional lung function measurements that are analogous to the routine tests of whole lung function (i.e. spirometry, lung volumes and gas transfer). The new techniques include lobar spirometry, lobar lung volume and local gas exchange.

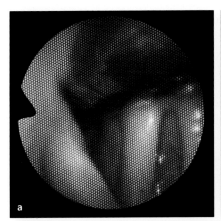
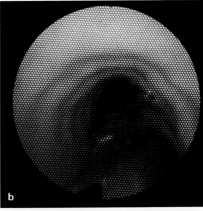

Fig. 5.4 Fibrescope positioned in the adult larynx (a) and trachea (b). A standard fibrescope (external diameter usually 5.7mm) may obstruct the larynx and reduce the tracheal diameter by 11% which results in a 17% increase in the FRC.

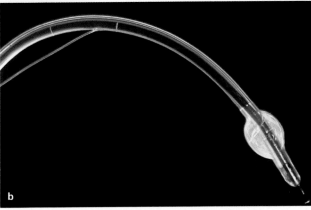

Fig. 5.5 Fibrescope in the endotracheal tube. The endotracheal tube (internal diameter 8mm) reduces the tracheal diameter by approximately 50% and a further reduction by 25% occurs when the fibrescope is inserted (a). The resistance to breathing is further increased by the length of the endotracheal tube which is obstructed (b).

Fibreoptic lobar spirometry

Fibreoptic lobar spirometry is necessarily different from its rigid predecessor because it is not possible to channel the gas from individual lungs or lobes through the smaller instrument. The solution to this seemingly difficult problem has been to use the fibreoptic bronchoscope to occlude the lobe or lung and then to study the effect of this temporary occlusion on whole lung function measured at the mouth.

Occlusion of a lobe or segment can be achieved with a balloon catheter, usually of the Swan-Ganz or Fogarty type, passed down the biopsy channel of the fibrescope. The size of the catheter that can be accommodated by the 2mm diameter channel of the standard instrument is 5 French (1.7mm) but a larger 7 French (2.3mm) catheter will pass down instruments with a 2.6mm diameter channel. The diameter of the inflated balloons of these catheters is 8.5mm and 11.5mm respectively and is sufficient to occlude lobar bronchi. However, they are not always large enough to block the main bronchi in adult patients. To achieve this it is necessary to insert a larger balloon catheter, such as a Foley, alongside the fibrescope. This can be guided under direct vision into the bronchus before it is inflated. A third method, which is not widely available, uses a modified fibrescope which has an inflatable cuff around the distal tip. With this it is possible to occlude any airway which can be entered.

The technique of fibreoptic lobar spirometry is not complicated. The bronchoscopy is performed transnasally and the spirogram or flow/volume loop is measured at the mouth while the fibrescope is resting in a major airway. The balloon catheter (Fig. 5.6) is then inflated in the appropriate lobe or segment, usually at FRC, and the spirometry is repeated. The functional contribution of the occluded portion is then estimated by the difference between the two values (see Fig. 5.6). It is important to remember that the

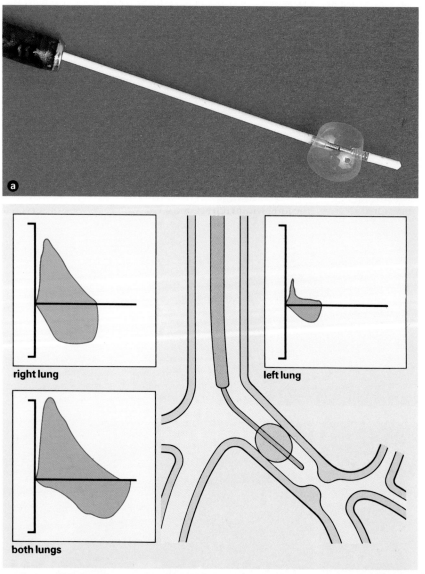

	right lung	left lung	both lungs
FEV₁ (litres)	2.2	0.3	2.6
FVC (litres)	2.5	1.3	3.7

Fig. 5.6 Balloon catheterization. View (a) shows the catheter inflated outside the fibrescope. After insertion of the fibrescope and measurement of the flow/volume loop, the balloon catheter is inflated and the measurements repeated. The diagram in (b) shows the results from a patient with stenosis of the left main bronchus. The contribution of the left lung when the right is occluded is small. The graphs represent volume (litres) on the abscissa against flow rate (litres/sec) on the ordinate.

FEV_1 and the forced vital capacity (FVC) may be reduced at the outset by the posture of the patient and the fibrescope itself. The value of lobar measurements is therefore best expressed as a percentage of optimum whole lung function.

In addition to single-breath lobar or bronchospirometry, the occlusion technique has also been used to predict the postoperative exercise capacity. In this instance, the ventilation ($\dot{V}E$) and oxygen uptake ($\dot{V}O_2$) of the patient are estimated during normal walking speed on a treadmill. Later the patient re-creates the $\dot{V}E$ and $\dot{V}O_2$ conditions of walking, during steady state exercise on a bicycle, with the fibreoptic bronchoscope in the airway. The relevant lobe is then occluded and postoperative exercise capacity is judged by the ability to maintain this $\dot{V}E$ and $\dot{V}O_2$ with the lobe excluded.

In some ways, occlusion bronchospirometry is better than its rigid predecessor because it gives information about the function of the lung that will remain after operation. There are, however, some fundamental disadvantages in common with traditional bronchospirometry. The temporary 'medical' lobectomy can never be perfect. Occlusion of a lobe may direct ventilation to other parts of the lung but will not remove the mechanical influences of the lobe and cannot predict with certainty how the rest of the lung will behave after surgery. The interference of an occluded lobe on the expiration from surrounding lung and the discomfort that it causes will be greatest if it is occluded at total lung capacity (TLC) and minimal if occluded at FRC. Also, occlusion will not abolish pulmonary blood flow and the resultant shunt will impair gas exchange. Pre-operative assessment of a patient with occlusion bronchospirometry will therefore produce a slightly greater functional deficit than will occur after the operation.

Lobar and segmental lung volumes measured at bronchoscopy

During life, the size of whole lungs can be measured by plethysmography, by radiography and by gas dilution methods. In reality, these methods measure different things; they estimate the gas content, the displacement volume and the accessible gas volume of the lungs respectively. In practice, there is good agreement between the methods in healthy lungs and in diseased lungs the difference between the methods is exploited to provide useful information (e.g. gas trapping). Unfortunately, it is difficult to measure the size of individual lobes and segments. Partial plethysmography is impossible and radiographic methods, including CT, can only be used if the fissures are visible. However, the fibreoptic bronchoscope has made a small contribution in this area by facilitating the measurement of lobar and segmental volumes by scintigraphic and gas dilution methods.

SCINTIGRAPHY — Lobar volumes can be measured by scintigraphy even though the boundary of the lobe may be invisible on plain radiography. It is possible to outline the lobe with radioactive gas and then measure the volume geometrically. This method, described by Williams and co-workers (1979), employs the introduction of a stream of krypton 81m (half-life thirteen seconds) into the lobe during inspiration via the channel of the fibrescope. Orthogonal gamma-camera images of the chest are taken to record the outlined lobe (Fig. 5.7). Once the image is magnified to normal size, the lobar dimensions are used to calculate volume using the same method that is employed for plain radiography when the fissures are visible.

GAS DILUTION — The measurement of lobar volume by gas dilution makes use of the fact that in normal lungs inspired gas mixes quickly and reaches more than ninety per cent of the lung within ten seconds. This process can be observed if an insoluble tracer gas such as helium or argon is added to the inspirate and the mixed concentration of the tracer is measured after expiration. The volume of the lung that is invaded can then be calculated and is known as the accessible volume (VA).

The measurement of the VA of lobes and segments can be made using the fibrescope to sample local expired gas.

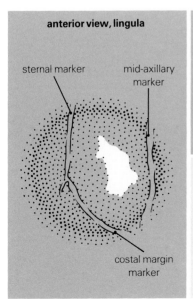

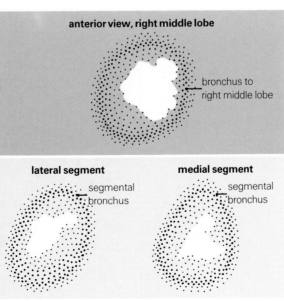

Fig. 5.7 Scintigraphic lobar volume estimation. Radioactive krypton 81m is introduced via the appropriate bronchi, as labelled, to outline lobes or segments, here shown in yellow. Once the boundaries are identified in two planes, the volume can be calculated. The diagrammatic examples shown are anterior views of the lingula and right middle lobe and segments.

Unfortunately, it is not possible to collect all the expired gas with the fibrescope; instead, the expired gas concentration is measured continuously. An example is shown in Figure 5.8 where a 20ml volume of argon is injected into the orifice of a lobe at the beginning of inspiration. The expired gas concentration measured by mass spectrometry shows a rise in tracer concentration to a plateau of alveolar gas. The volume of the lobe is calculated from the height of the plateau.

This inert gas bolus method compares well with scintigraphic, radiological and anatomical measurement of lobar volume. However, this is only true in the absence of airway obstruction. When gas mixing is poor, the expired gas concentration fails to reach a steady plateau and the VA cannot be calculated. Attempts to overcome this difficulty with an artificial rebreathing technique using a syringe to mix gas within an occluded lobe have proved disappointing. Gas dilution measurements of lobar volume can therefore only be made in the absence of airway obstruction.

Bronchoscopic measurements of local gas exchange
The characteristics of regional gas exchange can be assessed by analysis of local expired gas concentrations. The gases which are useful to study are the naturally exchanging metabolic gases, oxygen and carbon dioxide, and also alien inert gases which are introduced into the breath. For regional analysis the expired gases must be measured locally and the fibrescope is the ideal instru-

ment to position a sampling catheter in the required orifice. Gas samples taken in this way can be analysed as aliquots but more valuable data is obtained if the sampling of expired gas is continuous. This requirement places limitations on the type of gas analyser but the respiratory mass spectrometer is uniquely suited to endobronchial gas analysis.

The major advantage of the mass spectrometer is the ability to analyse rapidly several gases at once with a sample rate that is so low (20ml/min) that it will not on its own interfere with gas exchange. The fine disposable polyethylene probe of the mass spectrometer is only 1mm in diameter and can easily pass down the fibreoptic bronchoscope (Fig. 5.9). The concentrations of expired gases are measured easily by placing the probe beyond the tip of the scope within the lobe or segment under study. Concentration profiles are then obtained of the metabolic gases during tidal breathing or the pattern of inert gas behaviour during contrived manoeuvres.

Local metabolic gas exchange
Measurement of the exchange of the metabolic gases at the mouth, especially carbon dioxide, allows the assessment of how appropriate alveolar ventilation is to metabolic need. When these exchanges are measured within the lung they reflect the local relationship of ventilation to perfusion. The concentrations of oxygen and carbon dioxide measured by mass spectrometry at broncho-

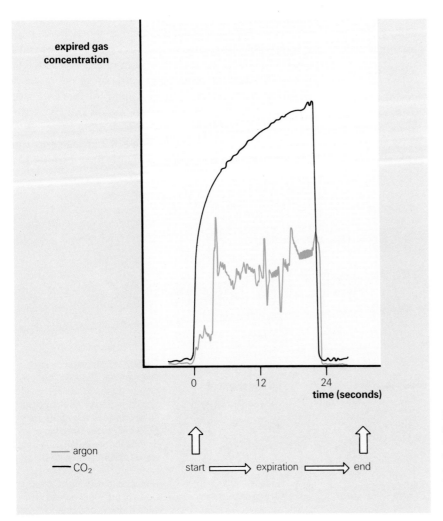

Fig. 5.8 Lobar volume estimation using an argon tracer. A small volume of tracer gas (20ml) is injected into a lobe at the beginning of inspiration. The expired gas concentrations are measured by mass spectrometry and the volume of the lobe is calculated from the height of the concentration plateau of the argon tracer gas.

scopy can supply this information. Normally, during tidal breathing, expiration is accompanied by a fall in the expired concentration of oxygen and a rise in the concentration of carbon dioxide as the sampled gas approaches mixed venous levels. Abnormalities in the shape and magnitude of these traces can be attributed to defects of gas mixing and to changes in local $\dot{V}/\dot{Q}$. Several patterns of tidal oxygen and carbon dioxide concentrations have been described (Fig. 5.10), resulting from bronchial or vascular occlusions. Although such measurements can be valuable, they are relatively insensitive since they may be influenced by factors other than gas mixing and $\dot{V}/\dot{Q}$.

Continuing gas exchange and changes in lung volume make carbon dioxide traces from different parts of the lung difficult to compare. For these reasons the introduction of inert gas markers of ventilation and blood flow, which are studied during the course of a single breath, provide more accurate information. However, it is often useful to combine tests using inert gases with measurements of oxygen and carbon dioxide concentrations.

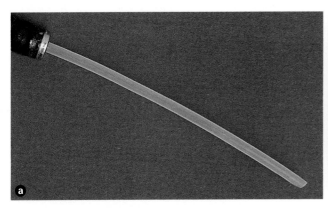
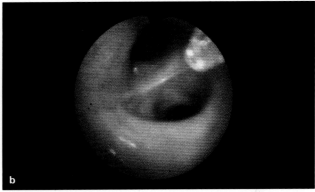

Fig. 5.9 Mass spectrometer probe used in fibreoptic bronchoscopy. The probe shown in (a), which is used to measure the concentration of expired gases, is only 1mm in diameter and passes easily down the fibreoptic bronchoscope. Once inside the bronchial tree (b) the probe can be positioned in any orifice.

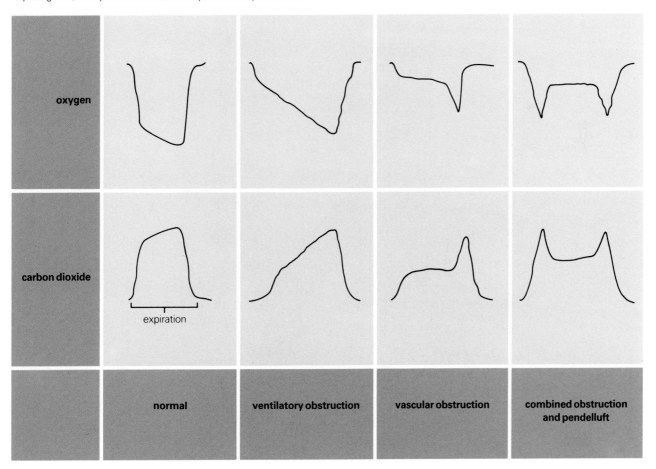

Fig. 5.10 Representation of several patterns of local metabolic gas profiles during expiration. Carbon dioxide and oxygen profiles are shown in a normal subject, during ventilatory obstruction, vascular obstruction and combined obstruction and pendelluft, where expired gas from other parts of the lung enters the affected area. The graphs represent time on the abscissa against expired gas concentration on the ordinate.

Bronchoscopic inert gas techniques – the single breath argon/freon test

The information that can be obtained from the study of inhaled alien gases depends upon the physical properties of the gases themselves and the ease with which they can be measured. Most of the gases which are currently used for physiological measurement are biologically inert and their usefulness depends upon their solubility. For example, helium and argon are gases which are insoluble in body fluids and will therefore mark the distribution of ventilation; nitrous oxide, acetylene and freon are gases which are partially soluble in tissue fluids and blood and the disappearance of these inspired gases is dependent upon their removal by pulmonary blood flow; gases which are highly soluble, for example, ether, will dissolve rapidly in all available water within the lung. Carbon monoxide is not an inert gas but it is used in the measurement of whole lung function because it is avidly removed by accessible haemoglobin. Its concentration can be measured by sampling in the same way as for the other gases.

For practical purposes the gases which are used for measurement are selected for the ease with which they can be measured as well as their physical properties.

Using mass spectrometry, argon is preferred to helium as a marker of ventilation because the mass number of helium is very low and difficult to measure. Freon 22 is preferred to nitrous oxide (N_2O) because of its lack of anaesthetic effect and also because the mass number of nitrous oxide is the same as carbon dioxide and cannot be distinguished unless an isotope is used. Similarly, the use of the mass spectrometer to follow expired carbon monoxide concentrations is limited because of mass number confusion with nitrogen, but an isotope ($C^{13}O^{18}$) can be used if it is desired. In spite of these practical difficulties, a combination single-breath test using a mixture of argon and freon to mark ventilation and perfusion has proved informative and has been adapted by Denison and others for the bronchoscopic measurement of regional lung function. The rest of this section is devoted to a description of the principles and practice of this method.

The principle of the single breath argon/freon test depends upon the continuous and simultaneous recording of the expired concentrations of argon and freon following the inhalation of a single breath of gas mixture. A typical trace is shown in Figure 5.11. During the

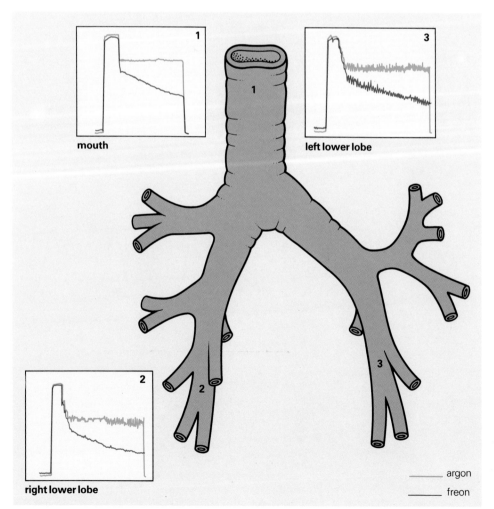

mouth

left lower lobe

right lower lobe

——— argon
——— freon

Fig. 5.11 Diagram illustrating single breaths labelled with an argon/freon mixture in a normal subject. Concentrations of the gases are measured during a single exhalation at the mouth and within the bronchial tree. The appearance of the traces is similar in each case. The recording positions and the corresponding traces are numbered.

beginning of the expiration, dead space gas is returned with concentrations of argon and freon similar to the inspired mixture. With the appearance of alveolar gas the concentration of argon falls to a plateau whose slope indicates the quality of gas mixing. It will be horizontal in normal subjects. The freon on the other hand separates slightly from the argon and thereafter continues to diverge as the expired concentration becomes lower as it is removed from the lung by the circulation.

Abnormalities of gas mixing and invasion of the lung by the argon will be manifest as the appearance of a negative slope to the expired trace. Reduction of perfusion, relative to ventilation, will be visible as a failure of separation of freon from argon and a reduction in the rate of freon disappearance.

If this test is applied to the whole lung by performing the test at the mouth, two important measurements can be made: first, the accessible volume (VA) can be calculated from the expired argon concentration; secondly, the rate of disappearance of freon is a measure of effective pulmonary blood flow, i.e. the flow of blood through that part of the pulmonary circulation that is in contact with the gas exchanging surface ($\dot{Q}p$). Flow through intrapulmonary shunts will not be detected. Using the fibreoptic bronchoscope to sample different parts of the lung, similar traces can be obtained (see Fig. 5.11), but unless the volumes of the subtended unit of lung are known, the same measurements of VA and $\dot{Q}p$ cannot be made. However, it is possible to calculate useful indices of gas mixing and blood flow.

Practical details

The aim of the test is to obtain single breath argon/freon profiles from each of the lobes of the lung. Occasionally it may be useful to perform segmental studies as well. The bronchoscopy itself is conventional and should, if possible, be performed transnasally with the patient in the supine posture (Fig. 5.12).

Before the fibrescope is inserted, the patient is asked to rest for several minutes to settle the circulation while the mass spectrometer and other equipment is prepared and calibrated. Following this, several single-breath tests are performed at the mouth, before the fibrescope is inserted, for two reasons: first, to familiarize the patient with the routine and secondly to measure whole lung VA and $\dot{Q}p$. The test is performed by asking the patient to empty the lungs completely then inhale sharply from a bag containing a mixture of argon, freon, oxygen and nitrogen (10% argon, 3.5% freon, 35% oxygen). When the lungs are full the patient holds breath for a few seconds and then breathes out slowly and steadily over twenty to thirty seconds, or as slowly as can be managed, until the lungs are empty once more. The concentration of expired gas is monitored continuously. Once the patient has mastered the technique, the probe is transferred to the fibrescope which is passed in the usual way and when it is in position similar traces can be obtained from each lobe after the gas is inhaled from the mouth. Usually it is necessary to leave an interval of approximately two to three minutes between tests to allow for gas equilibration. A full lobar survey therefore takes approximately twenty-five minutes. The skill required by the operator is relatively slight and only a little practice is necessary to keep the probe in the centre of the lobar orifice as it moves in expiration. Blockage of the probe by secretions is sometimes a problem but can be dealt with by syringe expulsion or by exchanging the probe.

Analysis of argon/freon traces

There are several ways to extract information from the argon/freon traces. They range in complexity from simple observation and shape recognition by eye to elaborate computer-dependent routines which make assumptions in an effort to extract more exact information. The current climate of opinion favours the analytic description of traces or the use of a simpler mathematical technique which is independent of major assumptions. Often, changes can easily be detected by eye especially when the area of study is distinctly different from other parts of the lung.

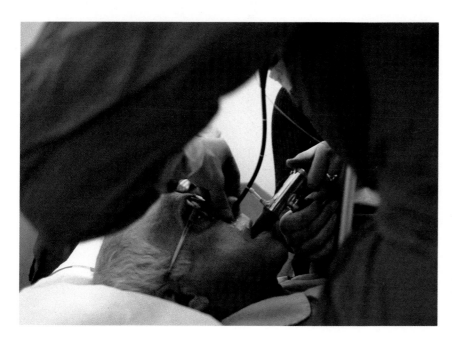

Fig. 5.12 Performing an argon/freon test. For this test the bronchoscopy is carried out transnasally with the patient supine. When the probe is in position the patient makes a full inspiration of inert gas mixture from the mouth.

In earlier computerized analyses of the argon/freon traces, corrections were required for normalization, realignment, reduction in lung volume and the effect of lung water. Consequently, results with these corrections were not often credible and it is better to consider the traces simply as records of entry into and removal of gas from the lungs. Denison has developed this observation to derive indices of the effectiveness of invasion (Ei), of gas mixing (Sm) and gas removal (Er). In this way, values from traces can be numerically compared to normal subjects or to other parts of the lung without implying a functional cause.

These semi-quantitative indices are obtained from the traces as shown in Figure 5.13. A box is drawn around the trace with the maximum argon concentration at the top boundary and zero at the bottom. The lateral boundaries are determined by the beginning and end of expiration. The Ei is the area under the insoluble gas trace expressed as a percentage of total box area. The temporo-spatial evenness (Sm) of gas mixing is expressed by convention as the slope of the argon trace between the twelfth and twenty-fourth second as a percentage of maximun argon concentration. This does of course make an assumption that the rate of expiration is even. Similarly the efficiency of removal (Er) is calculated from the slope of the freon trace over the same period.

Normal values are available to compare the indices between and within subjects. In this way useful semi-quantitative estimates of regional ventilation and perfusion can be obtained bronchoscopically. Unfortunately absolute measurements of lobar volume cannot be obtained simultaneously. Even so, important information can be gathered with this test.

Case Study 1

A forty-two-year-old woman developed dyspnoea after an aeroplane flight to Australia. She did not seek medical advice immediately but presented some weeks later when her symptoms persisted. She was a smoker but her plain chest radiograph was normal and routine lung function tests demonstrated a mild degree of airway obstruction with an unexpectedly low carbon monoxide transfer factor. After prolonged investigation she finally had a pulmonary angiogram (Fig. 5.14) which showed a discrete stenosis of the right main pulmonary artery which was later considered to be the consequence of an untreated pulmonary embolus.

Bronchoscopic argon/freon tests were performed prior to corrective surgery and repeated afterwards to judge its success. The pre-operative traces (Fig. 5.15) demonstrated a mild uniform abnormality of gas mixing (slope to argon trace) compatible with mild airway obstruction. The freon traces from all lobes of the right lung showed failure of separation and divergence indicating poor perfusion. By contrast, the effective blood flow to the left lung was very large. After surgery (Fig. 5.16) the overall appearance at the mouth was unchanged but blood flow returned to the right lung and was reduced in the left. The surgery achieved functional correction of the defect.

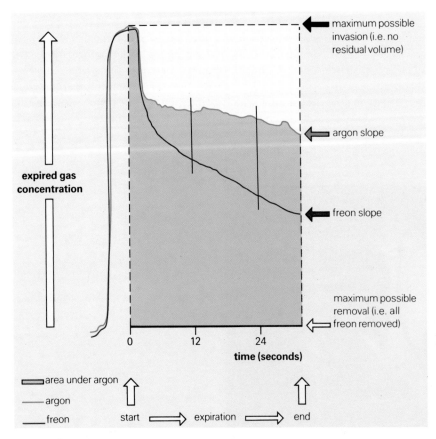
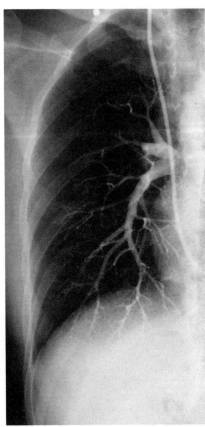

Fig. 5.13 Semi-quantitative analysis of the argon/freon trace. The analysis obtains indices of the effectiveness of gas invasion (Ei), of gas mixing (Sm) and gas removal or perfusion (Er). (See text for explanation.)

Fig. 5.14 Case study 1. Pulmonary angiogram showing discrete stenosis of the right main pulmonary artery immediately behind the heart (not easily seen in this view).

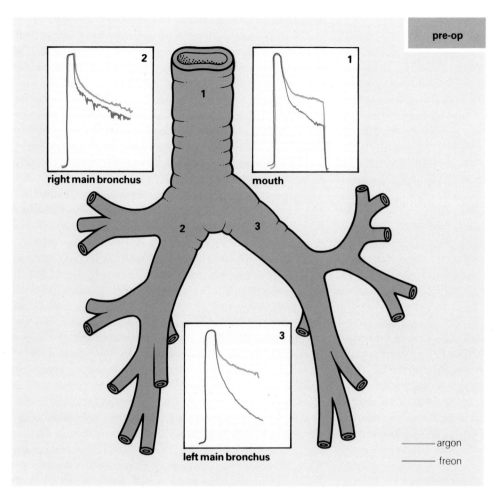

Fig. 5.15 Case study 1. Pre-operative argon/freon traces. A mild uniform abnormality of gas mixing is demonstrated by the slope of the argon trace, which is consistent with mild airway obstruction. Poor blood flow to the right lung is demonstrated by failure of separation of the freon trace. By contrast, blood flow to the left lung is exaggerated.

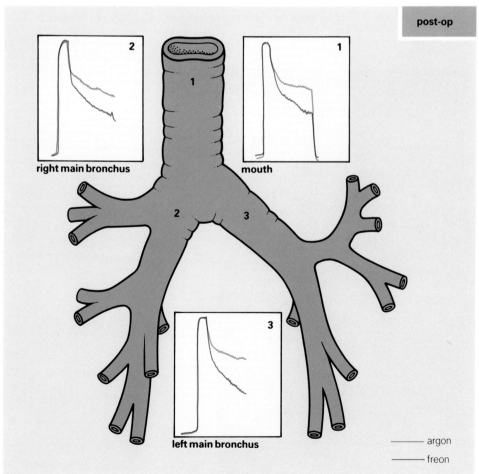

Fig. 5.16 Case study 1. Postoperative argon/freon traces. Surgery has achieved functional correction of the vascular obstruction to the right lung, indicated by an increase in blood flow to this lung. Blood flow to the left lung is proportionally diminished.

Case Study 2

A twenty-eight-year-old patient was a serving soldier with no symptoms of disease. He was referred for investigation of a routine chest radiograph which displayed an abnormality in the right hemithorax (Fig. 5.17). Routine investigations were normal but since his military career depended upon a satisfactory explanation of the abnormality he eventually underwent angiography which demonstrated a large arteriovenous malformation within the lung which fed from the systemic circulation.

Since the lesion was discrete, it was removed by surgery and bronchoscopic argon/freon tests were performed to assess improvement. The traces prior to operation (Fig. 5.18) were nearly normal except for reduced perfusion in the whole of the right lung. After surgery (Fig. 5.19) there was no improvement in the freon traces presumably because, unlike *Case Study 1*, there was no diversion of blood to the pulmonary capillary circulation. At the time of the second bronchoscopy he still had a small postoperative effusion and this minor effect on ventilation can be identified in the argon traces in the right lung.

Indications for Bronchoscopic Tests of Lung Function

The routine use of endobronchial tests has not been popular because they are invasive; other tests are preferred. However, most patients with lung disorders undergo bronchoscopy in the course of diagnosis and the additional measurements add little to the discomfort. In addition, the information obtained from the bronchoscopic measurement is complementary to that obtained from isotope or radiological studies. The major indications for clinical endobronchial studies relate to surgery. They include the prediction of postoperative lung function, the assessment of necessity for surgery and occasionally, appreciation of the results of surgery.

Prediction of postoperative lung function is most often required in those patients with operable lung cancer, but borderline whole lung function. For pneumonectomy, functional prediction can usually be made by perfusion scanning, but post-lobectomy function can only be estimated by endobronchial means. The relationship between pre-operative prediction and postoperative function has not been widely studied but two recent series by Bagg and by Corris have shown that both fibreoptic bronchospirometry and the bronchoscopic single-breath test can accurately predict the functional outcome of lobectomy.

The need for surgery in most conditions is determined by factors other than lung function. However, in some diseases there is a pathophysiological disorder which can be corrected by surgery. Examples include major airway obstruction, recurrent pleural effusion and bullous lung disease. In the latter, the functional contribution of the bulla and the consequences of its removal must be made prior to operation. A decade ago bronchoscopic lung function tests were the best available predictive investigations, but the development of CT has allowed the extent and physiology of bullous disease to be appraised. In this situation, bronchoscopic tests have now been overtaken as a primary investigation but can still provide useful information. CT has demonstrated that most bullae do not ventilate in the course of a single breath and they will not therefore be identified by the single breath bronchoscopic tests which only characterize lung that is accessible. Under these circumstances the bronchoscopic tests can provide a functional description of lung surrounding the bulla which helps to set it in its functional context. Examples of the combined use of CT and bronchoscopic tests of lung function are shown in *Case Studies 3 to 5*.

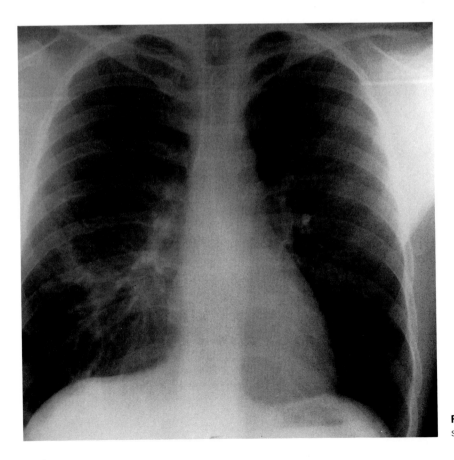

Fig. 5.17 Case study 2. Chest radiograph showing an abnormality in the right hemithorax.

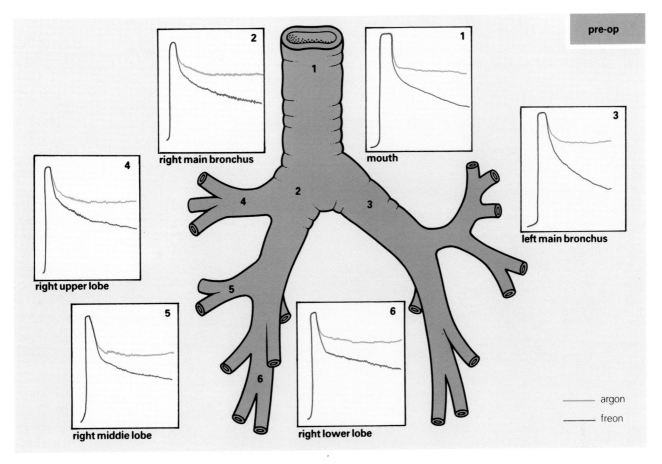

Fig. 5.18 Case study 2. Pre-operative argon/freon traces. The traces are normal except for reduced perfusion in the right lung due to a lesion in the right hemithorax.

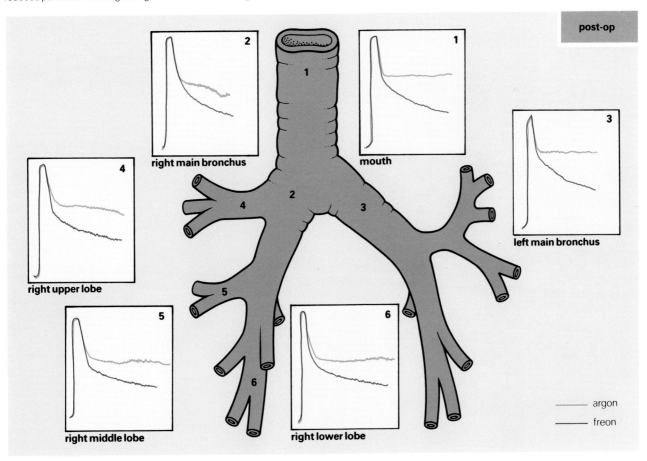

Fig. 5.19 Case study 2. Post-operative argon/freon traces. The lesion in the right hemithorax has been removed but there is no improvement in the freon trace.

Case Study 3

A policeman from the Middle-East complained of dyspnoea on exercise. As a young boy he had lobar pneumonia and his current chest radiograph (Fig. 5.20) suggested hyperlucency of the right upper zone. Routine lung function tests were surprisingly unhelpful except that there was 1.3 litres of gas trapping (TLC-VA). The bronchoscopic argon/freon tests (Fig. 5.21) confirmed that overall lung function appeared to be normal, except that the traces were a little short. This was also true in all lobes that were studied

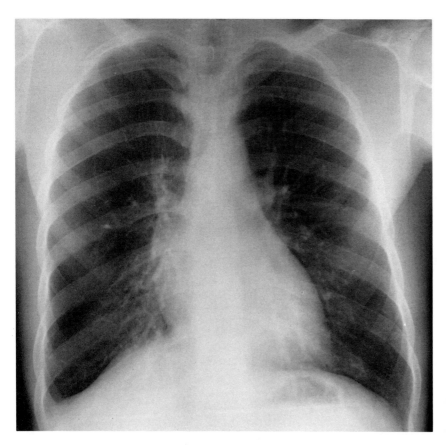

Fig. 5.20 Case study 3. Chest radiograph suggesting hyperlucency of the right upper zone.

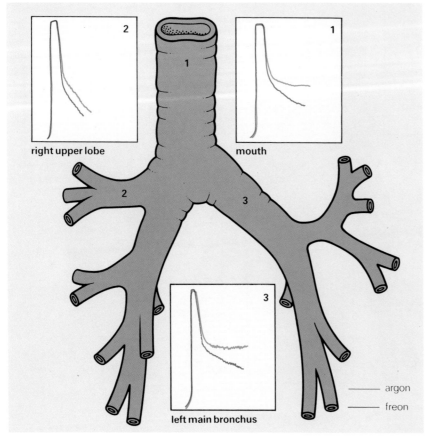

Fig. 5.21 Case study 3. Argon/freon traces. The right upper lobe trace shows a gross abnormality of ventilation and no perfusion.

except for the right upper lobe which had a gross abnormality of ventilation and no perfusion. Since this abnormality made no impact on the mouth trace it can be assumed that the lobe did not contribute to tidal ventilation. This was confirmed by a CT scan (Fig. 5.22) which looked almost normal in inspiration but highlighted the defective emptying of an emphysematous right upper lobe. This example illustrates the value of combined CT and endobronchial examination to assess the structural and functional impact of an abnormality.

Case Study 4

A fifty-seven-year-old patient presented with exertional dyspnoea of two years duration. He was a heavy smoker but had only occasional cough with scanty sputum. He had no other complaint and the past history included only chickenpox in early adulthood. A chest radiograph (Fig. 5.23) showed a large left-sided bulla extending the whole length of the thoracic cavity and compressing the upper lobe and lingula. The right lung fields appeared relatively normal except for calcification following resolution of chickenpox

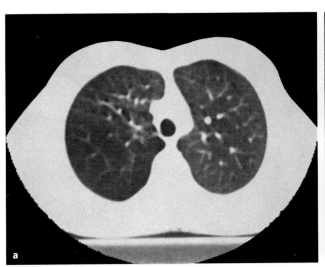 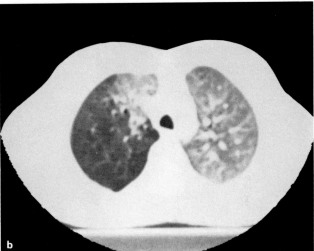

Fig. 5.22 Case study 3. The CT scan taken during inspiration (a) looks almost normal, but the scan taken during expiration (b) shows the defective emptying of an emphysematous right upper lobe.

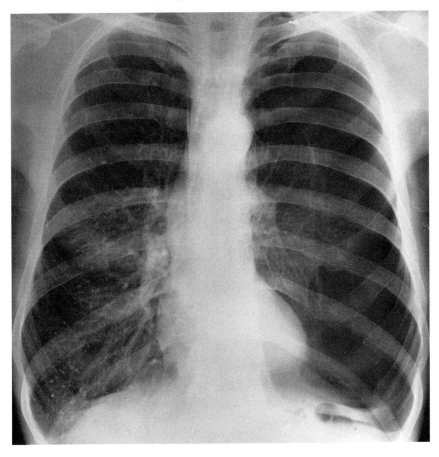

Fig. 5.23 Case study 4. Chest radiograph showing a large left-sided bulla which extends the whole length of the thoracic cavity and compresses the upper lobe and lingula.

pneumonia. A CT scan (Fig. 5.24) confirmed that there was a large bulla of 3,600ml in the left lung which did not change in volume during a vital capacity. In the expiratory scan, the mediastinum was displaced to the opposite side. The right lung, apparently normal on a routine chest radiograph, also had some small bullae anteriorly and posteriorly.

Argon/freon tests were performed (Fig. 5.25). The mouth trace showed an overall abnormality of ventilation and poor perfusion. The mouth trace was derived from the combination of the two lower lobe traces where the right lung had better preserved function than the left. Therefore neither lung was normal, but the bulla had adversely affected the ventilated lung surrounding it. Routine lung function tests (Fig. 5.26) showed the patient to have obstructive spirometry, increased lung volumes, reduced gas transfer and mild hypoxaemia.

In the light of these investigations the patient had a thoracotomy with plication of the bulla. He made a rapid postoperative recovery and one year later was symptom free. The postoperative lung function tests (see Fig. 5.26) showed an improvement in spirometry and a reduction in residual volume (RV) and FRC. Airways resistance (AWR) fell whilst the KCO was unchanged because the remaining lung was emphysematous.

Case Study 5

A thirty-year-old woman presented with haemoptysis. As a child she had bilateral pulmonary tuberculosis with extensive residual damage and cyst formation in the left lung (Fig. 5.27). The cause of the haemoptysis was visible on CT scanning (Fig. 5.28). A supine scan confirmed the presence of a large cyst in the upper lobe and a smaller cavity in the apex of the lower

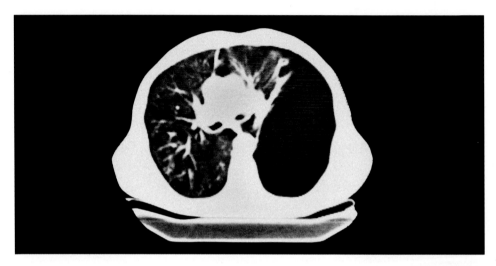

Fig. 5.24 Case study 4. A CT scan confirms the presence of a bulla which is compressing lung tissue anteriorly and displacing the mediastinum.

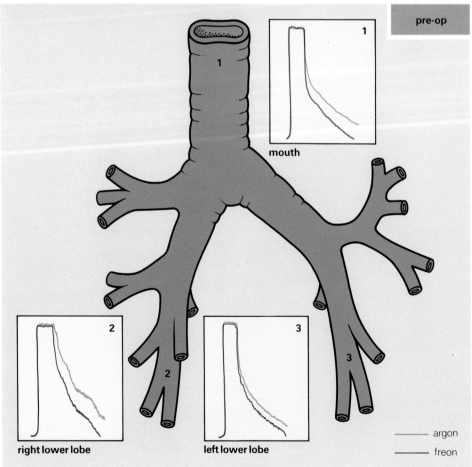

pre-op

mouth

right lower lobe

left lower lobe

—— argon
—— freon

Fig. 5.25 Case study 4. Pre-operative argon/freon traces. The mouth trace shows an overall abnormality of ventilation and poor perfusion.

Pulmonary Function Tests

	FEV₁ (litres)	FVC (litres)	FEV₁/FVC	TLC (litres)	RV (litres)	FRC (litres)	AWR (kPa/l/s)	DLCO (mmol/min/kPa)	KCO (mmol/min/kPa/l)	PaO₂ (kPa)	PaCO₂ (kPa)
predicted	3.3	4.4	70%	6.7	2.4	3.2	<0.2	11·0	1·44	–	–
pre-op	0.99	2.43	40%	7.42	4.67	6.17	0·98	4·40	1·04	8·7	4·5
post-op	1.92	4.34	44%	7.53	3.33	5.28	0·29	6·03	1·02	10·7	5·1

Fig. 5.26 Case study 4. Pulmonary function tests before and after thoracotomy for a left-sided bulla which extended the whole length of the thoracic cavity.

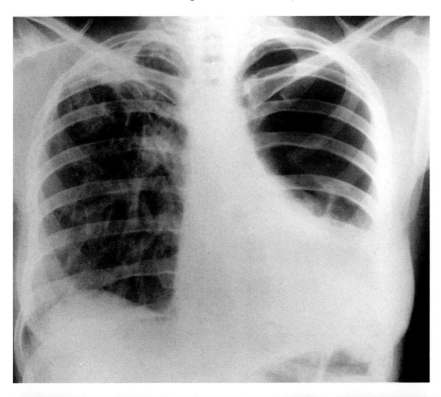

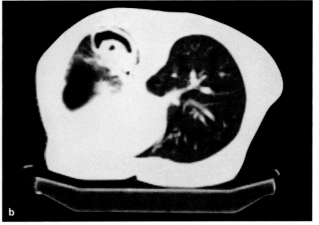

Fig. 5.27 Case study 5. Chest radiograph showing cyst formation in the left lung due to pulmonary tuberculosis.

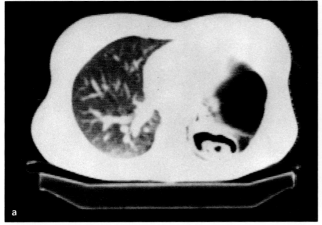

Fig. 5.28 Case study 5. (a) prone and (b) supine CT scans which confirm the presence of a large cyst in the upper lobe and a smaller cavity containing an aspergilloma.

lobe which contained an aspergilloma. When the patient was turned to the prone position the fungus ball fell forwards in the cavity. Bronchoscopic tests (Fig. 5.29) were performed to assess whether her symptoms could be resolved by left pneumonectomy without the loss of significant functioning lung. The mouth traces showed a normal pattern of gas mixing and perfusion which were also seen in the right main bronchus. The traces from the left lung, however, were grossly abnormal but made no impact on the mouth trace. The left lung therefore could make no useful contribution to overall performance and could be safely removed.

Endobronchial Investigation – Research and Future Developments

The distinction between clinical endobronchial tests and research studies should not be too great. All of the clinically useful tests have been derived from studies which were originally designed to gain greater knowledge of respiratory physiology. Some of the original bronchospirometry investigations identified the postural and gravitational changes of ventilation and blood flow before they could be confirmed by other means. The development of the fibreoptic bronchoscope has increased the anatomical range of measurement and reduced the experimental interference on the subject. Unfortunately the trade-off to these advantages has been the confinement of sampling through a small channel. Until now it has only been possible to measure gas concentrations, pressure and temperature directly, while balloon occlusion can only be applied to small airways. The introduction of miniaturized measuring devices is likely to reduce these limitations. The development of catheter tip pressure transducers, flow meters and other devices will revolutionize fibreoptic bronchoscopic measurement in the same way that mass spectrometry improved rigid bronchoscopy. Some of these instruments are already available and the future of endobronchial measurement with the fibreoptic bronchoscope will only be limited by the inventiveness of the researcher.

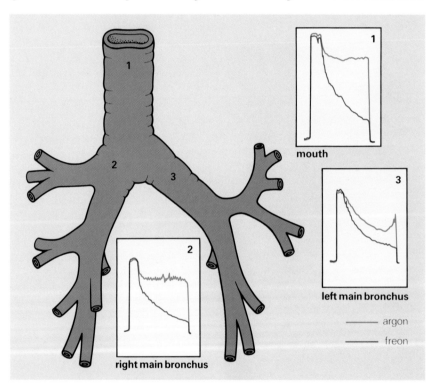

mouth

left main bronchus

— argon
— freon

right main bronchus

Fig. 5.29 Case study 5. Argon/freon traces. The grossly abnormal trace from the left lung makes no impact on the mouth trace which is normal. The left lung could therefore be removed without affecting overall lung performance.

REFERENCES

Albertini RE, Harrel JH, Moser KM (1975). Management of arterial hypoxemia induced by fiberoptic bronchoscopy. *Chest,* **67,** 134-136.

Bagg LR, Cox ID (1984). Balloon occlusion of the bronchi at fibreoptic bronchoscopy: application to physiological assessment before lung resection for bronchogenic carcinoma. *Thorax,* **39,** 236.

Brach BB, Escano GG, Harrell JH, Moser KM (1976). Ventilation-perfusion alterations induced by fiberoptic bronchoscopy. *Chest,* **69,** 335-337.

Corris PA, Kendrick AM, Gibson GJ (1985). Use of bronchoscopic single-breath tests to predict functional results of lobectomy. *Thorax,* **40,** 236.

Denison DM, Waller JF (1982). Interpreting the results of regional single-breath studies from the patient's point of view. *Bulletin Europeen de Physiopathologie Respiratoire,* **18,** 339-351.

Dubrawsky C, Awe RJ, Jenkins DE (1975). The effect of broncho-fiberscopic examination on oxygenation status. *Chest,* **67,** 137-140.

Hugh-Jones P, West JB (1960). Detection of bronchial and arterial obstruction by continuous gas analysis from individual lobes and segments of the lung. *Thorax,* **15,** 154-164.

Katz AS, Michelson EL, Stawicki J, Holford FD (1981). Cardiac arrhythmias. Frequency during fiberoptic bronchoscopy and correlation with hypoxemia. *Archives of Internal Medicine,* **141,** 603-606.

Matsushima Y, Jones RL, King EG, Moysa G, Alton JDM (1984). Alteration in pulmonary mechanisms and gas exchange during routine fiberoptic bronchosopy. *Chest,* **86,** 184-188.

Pierce RJ, Pretto JJ, Rochford PD, McDonald CF, Manan JA, Barter CE (1986). Lobar occlusion in the pre-operative assessment of patients with lung cancer. *British Journal of Diseases of the Chest,* **80,** 27-36.

Pierson DJ, Iseman MD, Sutton FD, Zwillich CW, Creagh CE (1974). Arterial blood gas changes in fiberoptic bronchoscopy during mechanical ventilation. *Chest,* **66,** 495-497.

Shrader DL, Lakshminarayan S (1978). The effect of fiberoptic bronchoscopy on cardiac rhythm. *Chest,* **73,** 821-824.

West JB (1977). Regional Differences in the Lung. Academic Press, London: 488pp.

Williams SJ, Pierce RJ, Davies NJH, Denison DM (1979). Methods of studying lobar and segmental function of the lung in man. *British Journal of Diseases of the Chest,* **73,** 97-112.

6. Therapeutic Aspects of Bronchoscopy

M. R. Hetzel MD FRCP

INTRODUCTION

In recent years, the fibreoptic bronchoscope has been used more frequently as a therapeutic instrument. Despite the relatively small size of the instrument channel (2.0-2.6mm), the flexible bronchoscope can be useful in removing inspissated secretions and some foreign bodies; particularly the less rigid foreign bodies whose size allows their removal through the instrument channel.

However, some of the positive features of the fibreoptic instrument, such as its small diameter, which make it ideal for use under local anaesthetic, can limit its therapeutic application. For instance, if the foreign body is large (e.g. a peanut) rigid bronchoscopy is essential, and would be used instead of the fibreoptic bronchoscope. Alternatively, there are situations where a combined approach is the ideal technique.

It cannot be overemphasized that therapeutic bronchoscopy is more hazardous than routine diagnostic bronchoscopy; particularly with regard to haemorrhage. Availability of a thoracic surgeon or experience with the rigid instrument is therefore important when contemplating procedures such as removal of foreign bodies.

The limited therapeutic applications of bronchoscopy have recently been improved with the advent of high power (class IV) lasers, which can be transmitted down optical fibres passed through the fibrescope. Thus precise cutting and coagulation for airway surgery can be achieved on any part of the tracheobronchial tree which is within the range of the fibreoptic bronchoscope. Laser photoresection of tumours is already of proven value in palliation, and may possibly give improved survival and even cures in the future. A major part of this chapter is therefore devoted to laser techniques but the longer established therapeutic applications of bronchoscopy will first be discussed.

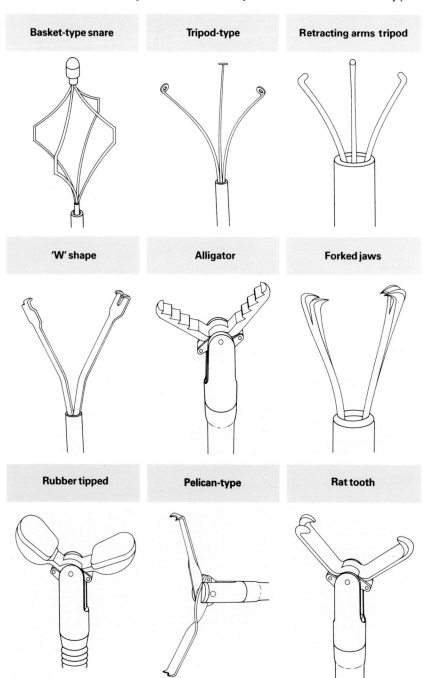

Basket-type snare   **Tripod-type**   **Retracting arms tripod**

'W' shape   **Alligator**   **Forked jaws**

Rubber tipped   **Pelican-type**   **Rat tooth**

Fig. 6.1 The flexible grasping instruments currently available for foreign body removal. In addition to those illustrated, magnetic extractors are also used. None of these can be withdrawn through the biopsy channel while holding a foreign body. (Olympus series: courtesy Key Med.)

REMOVAL OF FOREIGN BODIES

When a patient who has inhaled a foreign body presents with acute stridor and impending asphyxia immediately after inhalation, there is little difficulty in making the diagnosis. Occasionally, however, inhalation of a foreign body may not provoke sufficient symptoms to attract immediate medical attention and the episode may subsequently be forgotten. Thus, when investigating recurrent consolidation or collapse in the same lung or lobe and when investigating bronchiectasis of recent onset, the inhalation of a foreign body should be considered.

Prompt treatment usually leads to a good prognosis, but delay will result in irrevocable damage. High risk groups include children, alcoholics, epileptics and also elderly and demented patients. Some of these patients may have no recollection of accidental inhalation or they may be incapable of giving an adequate history. If the history or chest radiograph is indicative of a large foreign body, rigid bronchoscopy is advisable. This enables the use of biopsy forceps with large jaws or, preferably, purpose-built foreign body or 'peanut' forceps which will secure the object more readily. General anaesthesia will facilitate instrumentation, particularly if several attempts at removal are needed.

Foreign bodies may be encountered as a relatively unsuspected finding at fibreoptic bronchoscopy under local anaesthesia (see *Case Studies 2 and 3*). In this situation, it is reasonable to attempt removal, which may be successful for smaller foreign bodies. Ordinary fibreoptic biopsy forceps will rarely be capable of gripping a foreign body, but a variety of grasping instruments are available (Fig. 6.1). These include rubber-toothed forceps, magnetic extractors and snares. In using snares, the possibility of being unable to disentangle the snare from a foreign body, which is deeply embedded in the bronchial wall, should be taken into consideration.

If a foreign body is not easily removed but does not immediately threaten the patient's airway, it is important not to persist with the attempted removal. Repeated instrumentation will result in oedema or haemorrhage and cause more serious obstruction of the airway; rigid bronchoscopy should then be performed. Some foreign bodies are very difficult to remove with either instrument and may be partly buried in granulations. Surgical help should then be sought and bronchotomy may be required.

Case Study 1

An eighty-two-year-old man had a past history of inpatient treatment for schizophrenia and had been a heavy smoker all his life. He was unable to give any coherent history but his wife reported a one-month history of cough and breathlessness. Physical examination revealed signs of collapse of the right lower lobe, and the cause was found to be a radio-opaque foreign body on chest radiography (Fig. 6.2). At bronchoscopy this was subsequently found to be a gold crown which was very mobile in the bronchial tree and which was actually found in the left main bronchus. Because of its highly polished surface it was very difficult to grasp, but was successfully removed using rigid bronchoscopy. The right lower lobe subsequently re-expanded.

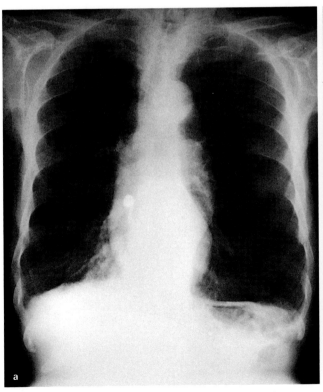
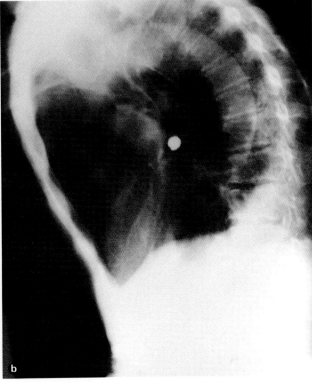

Fig. 6.2   Case study 1. (a) posteroanterior and (b) lateral chest radiographs taken prior to removal of a radio-opaque gold crown. The right lower lobe collapse seen here was caused by the foreign body. The gold crown can be better seen in the lateral view.

Case Study 2

A seventy-seven-year-old man had a history of heavy alcohol intake and presented with right-sided pleuritic pains. Chest radiography revealed right upper lobe shadowing (Fig. 6.3) and, since the patient had also smoked heavily, a provisional diagnosis of peripheral bronchial carcinoma was made. As there was no obvious evidence of metastatic disease, a transbronchial biopsy was attempted under radiographic screening to examine the histology of this lesion. The proximal bronchial tree was assessed for operability at the same time.

At fibreoptic bronchoscopy, a hard irregular mass was found in the orifice of the right intermediate bronchus (Fig. 6.4); however, it was not possible to inspect the lobar bronchi beyond it. Superficially it resembled a tumour but, as attempts were made to take a biopsy, it became clear that it was a foreign body embedded in granulation tissue. Although it could not be removed with fibreoptic instruments, it was successfully removed using rigid bronchoscopy with general anaesthesia. The foreign body was subsequently identified as a vertebra from a rabbit. The patient had no explanation for its presence in his bronchial tree, but had presumably inhaled it during one of his drinking bouts.

The diagnosis was therefore revised to a pneumonic process in the right upper lobe resulting from proximal obstruction by a foreign body. Paradoxically, the initial diagnosis proved to be correct. The peripheral shadow persisted and subsequent investigations indicated the presence of a squamous cell carcinoma; the unsuspected foreign body was an incidental finding.

Case Study 3

A fifty-nine-year-old woman was referred by the radiotherapy department with a two-year history of recurrent attacks of cough, pyrexia and purulent sputum and episodes of wheezing. Six years previously, she had undergone mastectomy and radiotherapy for carcinoma of the right breast. Several chest radiographs indicated episodes of consolidation and partial collapse in the left upper lobe. A fixed wheeze was audible over the left upper lobe. There was no history of asthma and respiratory function tests showed a mild restrictive defect, perhaps attributable to some pulmonary fibrosis from the previous radiotherapy.

A clinical diagnosis of obstruction of the left upper lobe bronchus, possibly by endobronchial metastasis from the breast carcinoma, was made and fibreoptic bronchoscopy performed. A hard irregular white lesion obscuring the orifice of the left upper lobe bronchus was found. This was successfully removed *in toto* by being grasped with flexible forceps through the fibrescope and then by removing the fibrescope completely while maintaining hold of this foreign body. On reintubation, stenosis of the left upper lobe bronchus was seen (Fig. 6.5). The foreign body was found to be amorphous calcified material on histology, and biopsies of the mucosa of the upper lobe bronchus showed chronic inflammatory changes only. Thus the foreign body was presumably a broncholith associated with previous invasion of the bronchial wall by a tuberculous node. (There was a small amount of calcification at the left hilum on chest radiography but this was unimpressive.) There was, however, no history of tuberculosis and the Heaf test was negative (grade 0). Follow-up bronchoscopies over the following year showed persistent stenosis of the left upper lobe, but the patient had no further symptoms.

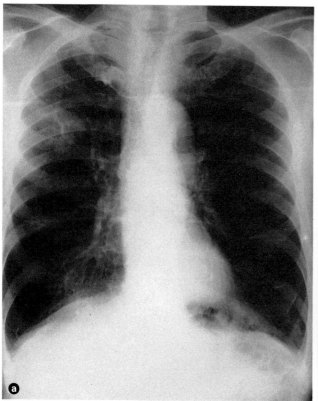

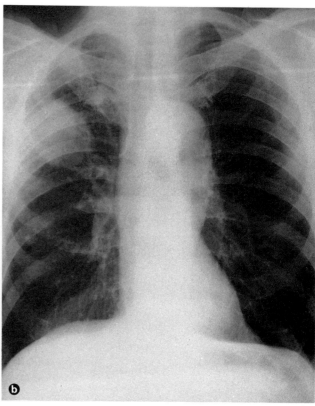

Fig. 6.3 Case study 2. View (a) was taken at presentation with a suspected tumour in the right upper lobe. One year later, view (b), the tumour has enlarged despite removal of the foreign body from the right intermediate bronchus.

The fibrescope alone can be effective in foreign body removal (see *Case Study 3*), particularly if they are small and present in the upper lobes where access is poor with rigid instruments. The foreign body will, however, almost invariably be too large to withdraw through the fibrescope instrument channel; thus the whole fibrescope has to be removed with it from the bronchial tree. It is usually easy to reintubate the patient during treatment under local anaesthesia; however, this will be poorly tolerated if performed repeatedly, as may be necessary if the foreign body breaks up and cannot be removed fully at the first attempt. Thus the combined use of the fibrescope through the rigid bronchoscope is the ideal method for smaller lesions in the upper lobes, since the fibrescope can then be removed and replaced with impunity.

Broncholiths frequently present with haemoptysis and are sometimes coughed up spontaneously. When planning their elective removal using bronchoscopy, precautions should be taken to deal with a possible haemorrhage.

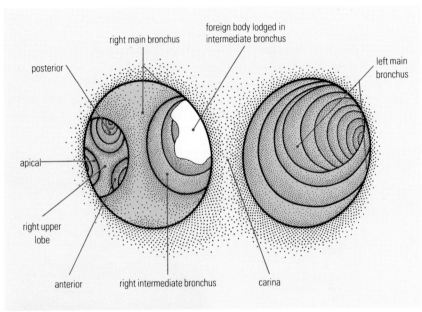

Fig. 6.4 Case study 2. Diagrammatic representation of bronchoscopic findings. A foreign body (rabbit vertebra), found in the intermediate bronchus at fibreoptic bronchoscopy for suspected bronchial carcinoma, required rigid bronchoscopy for successful removal.

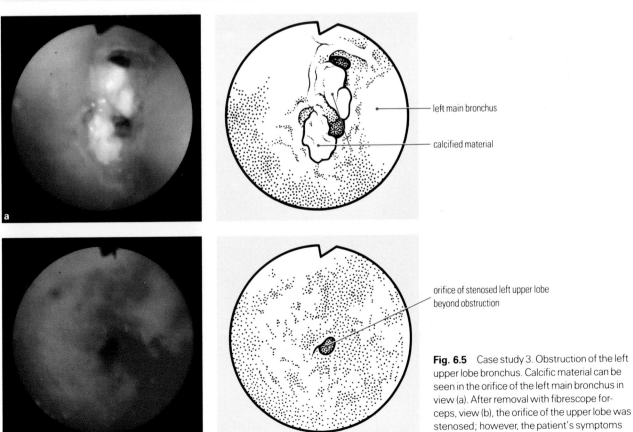

Fig. 6.5 Case study 3. Obstruction of the left upper lobe bronchus. Calcific material can be seen in the orifice of the left main bronchus in view (a). After removal with fibrescope forceps, view (b), the orifice of the upper lobe was stenosed; however, the patient's symptoms were cured by bronchoscopy.

BRONCHOSCOPY IN INTENSIVE CARE

The fibreoptic bronchoscope can be invaluable in intensive care, when managing patients with collapse of either a lung or lobe resulting from mucus plugs or a blood clot. By using a suitable adaptor (Fig. 6.6), the fibrescope can be passed through an endotracheal tube while maintaining an airtight seal; thus positive pressure ventilation (PPV) can be continued. Prolonged bronchoscopy is often possible using this system, but the presence of the fibrescope in the endotracheal tube will cause a considerable reduction in ventilation, particularly if suction is used constantly (see *Chapter 5*). Anaesthetic assistance is necessary to monitor and control ventilation so that the bronchoscopist can safely concentrate on the bronchoscopy. It may be necessary to repeatedly remove the fibrescope to allow the anaesthetist to ventilate the patient more effectively. Simple inspection of chest wall movement gives the earliest warning that the fibrescope is causing ventilatory failure, and action should be taken at this stage before circulatory signs of hypoxia and hypercapnia develop.

It is not usually necessary to instill local anaesthetic through the fibrescope in patients on ventilators, because they are already sedated and usually paralysed. It is important, however, to ensure that sedation is adequate, particularly in patients who have had a long period of ventilation and who may therefore be less heavily sedated.

Mucus plugs can be removed by suction and this is usually greatly facilitated by irrigation with saline. Excessive volumes of saline should be avoided, however, as this may worsen gas exchange for several hours following bronchoscopy; 20ml aliquots are adequate for this purpose. Viscid sputum can often be cleared by maintaining suction and withdrawing the tip of the fibrescope by 1-2cm. Wide-channel models have some advantage in this application.

Occasionally, biopsy forceps may be needed to remove hard plugs of mucus. The suction valves of fibreoptic bronchoscopes can become clogged with these viscid secretions. If suction becomes weak during the procedure, the valve should be dismantled and cleaned and the

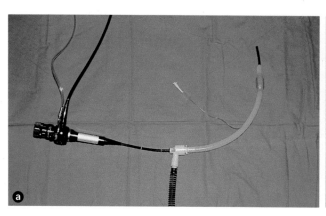

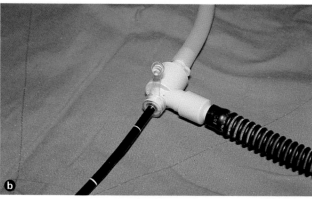

Fig. 6.6 A fibreoptic bronchoscope inside an endotracheal tube. This method is used for clearance of airways during mechanical ventilation. View (a) shows the fibrescope, endotracheal tube and catheter mount.

An enlarged view of the seal to prevent leakage of air around the fibrescope is shown in (b). When the fibrescope is removed a plug can be attached to the seal to allow continued ventilation.

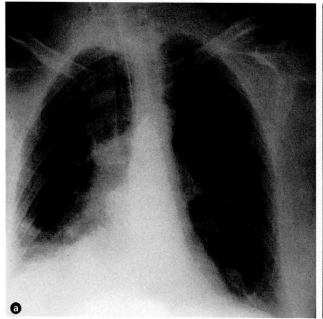

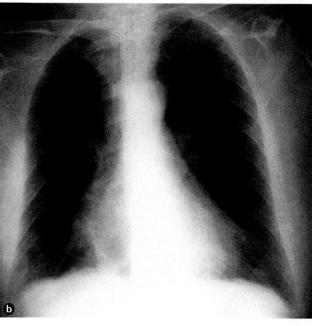

Fig. 6.7 Case study 4. Chest radiographs taken during mechanical ventilation for acute asthma. View (a) shows the collapse and consolidation of the right middle and lower lobe. View (b) was taken after bronchoscopy. The diaphragm and right heart border are visible again.

cleaning brush passed through the suction channel of the fibrescope. If this problem occurs during clearance of mucus plugs from a conscious patient examined under local anaesthesia, the fibrescope can be left *in situ* while this is performed.

Collapsed lung or lobes resulting from pulmonary haemorrhage may also be seen, caused by obstruction of the bronchus with a blood clot. Again significant haemorrhage may occur following removal of the clot. Also, the clot may have been protecting the remainder of the bronchial tree from obstruction by blood. If a large clot is occluding a main bronchus, it may not be easy to remove with the fibrescope alone; thus the merits of using a rigid bronchoscope should be carefully considered before attempting removal. The inexperienced bronchoscopist should appreciate that old blood clots can become very pale and are easily mistaken for tumours; their very smooth outline is usually helpful in identification.

Case Study 4

A sixty-nine-year-old woman had a history of life-long asthma and had been taking oral steroids for many years. At best, she had an exercise tolerance of one hundred metres on level ground, but managed to continue working as a shop assistant. She presented with an acute exacerbation of asthma and collapse of the right lower lobe from

mucus plugging. She gradually deteriorated over the next twelve days, despite physiotherapy, prednisolone (60mg per day) and bronchodilator drugs.

After six days of treatment she had become exhausted with persistent hypoxia and eventually developed hypercapnia; she was treated with mechanical ventilation. After forty-eight hours ventilation there was little improvement and her chest radiograph showed persistent collapse of the right lower lobe. The fibreoptic bronchoscope was therefore passed through the endotracheal tube and the presence of mucus plugs in the orifice of the right lower lobe bronchus was confirmed. A total of 300ml saline was used to remove the proximal plugs. Many smaller plugs were seen in the aspirate from the lower lobe. Re-expansion of the lower lobe subsequently occurred and she was weaned off the ventilator after a further forty-eight hours (Fig. 6.7). Significant improvement in arterial blood gases was also seen (Fig. 6.8).

Case Study 5

A twenty-eight-year-old asthmatic woman, who had fallen down some stairs, was admitted with severe left-sided chest pain from fractures of her ninth and tenth ribs (Fig. 6.9). This was noted to have produced a small flail area posteriorly. On admission, her asthma was stable. She was treated with intercostal block with bupivacaine to control her pain and minimize the need for opiates in view of her asthma. Despite these measures, together with vigorous bronchodilator therapy, her asthma

Fig. 6.8 Case study 4. Table showing the improvement in arterial blood gases following bronchial lavage in a patient with acute asthma.

Arterial Blood Gases Pre and Post Bronchial Lavage			
	pre-lavage	**post-lavage**	
		3 hours	30 hours
pH	7·44	7·31	7·39
PCO$_2$ (mm Hg)	36·0	41·9	31·5
PO$_2$ (mm Hg)	64·6	71·2	117·2
FIO$_2$	40%	20%	20%

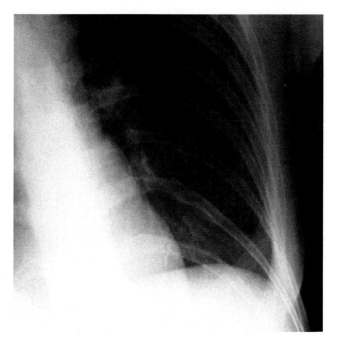

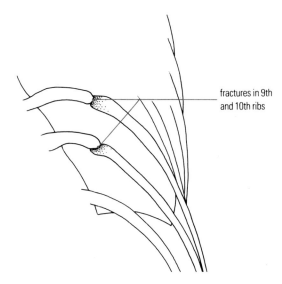

fractures in 9th and 10th ribs

Fig. 6.9 Case study 5. Radiograph showing appearance on admission with fractures of left ninth and tenth ribs.

worsened and after three days she developed collapse of both lower lobes (Fig. 6.10). She was treated with oral steroids and physiotherapy and the right lower lobe collapse resolved.

On the morning of the twelfth day of admission she awoke acutely breathless and was found to have a large left pneumothorax (Fig. 6.11), presumably from puncture of the lung by the fractured ribs. Insertion of an intercostal drain was not completely effective in relieving her breathlessness and she was noted to have a paradoxical shift of the mediastinum towards the left side. This was explained by persistent collapse of the left lower lobe. Fibreoptic bronchoscopy was therefore performed. Tenacious mucus plugs were removed from the left lower lobe. Subsequently the lower lobe re-expanded, her breathlessness improved and the intercostal drain was removed (Fig. 6.12).

Bronchoscopic removal of mucus plugs which have caused collapse is mandatory if physiotherapy, plus vigorous bronchodilator therapy and oral steroids, fails to re-expand the lung within a few days.

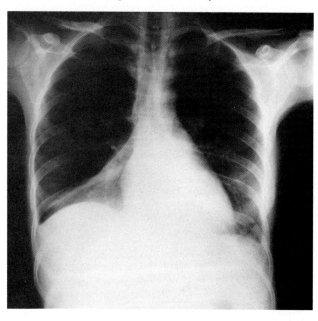

Fig. 6.10 Case study 5. Chest radiograph showing collapse of both lower lobes due to mucus plugs.

Bronchial lavage with carefully controlled volumes of saline has sometimes been effective in intractable acute asthma by removing mucus plugs from the smaller airways. Caution must be exercised since it is a potentially dangerous treatment. Gas exchange may worsen acutely; if the patient is not already on a ventilator, at least a short period of ventilatory assistance may be required after lavage. Nevertheless, if all else fails, bronchial lavage is a measure that merits consideration.

Finally, the fibrescope can sometimes be used to position an endotracheal tube when intubation is difficult; for example, in acute epiglottitis, in patients with tumours in the mouth or pharynx or when the neck cannot be fully extended as in, for example, ankylosing spondylitis. The tube is threaded over the fibrescope and pushed up to the control handle. Once the larynx has been located, the tube is then threaded down over the fibrescope and through the vocal cords. The nasal approach to bronchoscopy may still be used if an appropriately sized endotracheal tube (e.g. 6mm) is employed.

USE OF LASERS IN THERAPEUTIC BRONCHOSCOPY

Over the years there have been a number of attempts to develop effective means of performing airway surgery through the fibrescope while maintaining good haemostasis. This is particularly relevant in patients with inoperable conditions, especially those with tracheobronchial tumours. Previous work with cryoprobes or diathermy has had limited success. These probes need to be in physical contact with the tumour and finite periods are required for them to have an effect. Consequently, they can become stuck to the tumour and bleeding may occur on removal. The extent of thermal damage cannot be precisely controlled since, if the probes are misplaced, they can damage normal tissues as they cannot be instantly switched off. They are relatively bulky and have limited manoeuvrability.

Most of these problems have now been overcome using laser technology where there is no direct contact with the tumour. Lasers can be aimed precisely as a spot

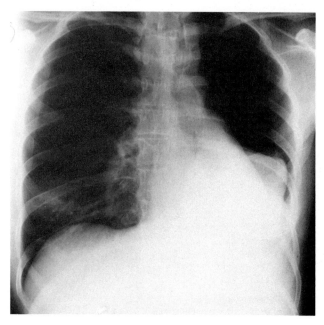

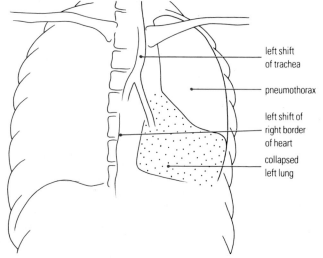

Fig. 6.11 Case study 5. Chest radiograph showing left pneumothorax with a paradoxical shift of the mediastinum to the left due to persistent mucus plugs obstructing the left lower lobe bronchus.

of approximately 1mm diameter, and if incorrectly aimed they will not damage normal tissues as they can be instantly switched off. The ideal lasers for bronchoscopic application need to be passed through optical fibres, which are guided down the fibrescope suction channel, providing a system which is as manoeuvrable as the fibreoptic bronchoscope itself.

Two main laser techniques are available for the treatment of tumours. First, direct thermal treatment uses high power (class IV) lasers which heat the target tissue to several hundred degrees celsius causing vaporization. Lower exposures carbonize the tissue, which can then be removed with biopsy forceps. The heat also shrinks blood vessels achieving haemostasis. These effects are termed photoresection and photocoagulation respectively. Tumour cells will die after modest rises in temperature of approximately 8°C. The eventual area of cell death will therefore be larger than the immediately visible laser burn.

Secondly, in photodynamic therapy, laser light at low powers is used to excite acid derivatives of haematoporphyrin, haematoporphyrin derivative (HPD) or, in a purer form, dihaematoporphyrin ether (DHE), which are preferentially retained in tumour cells after intravenous administration. When activated, these agents induce a cytotoxic effect through formation of singlet oxygen.

The applications of these two techniques will be discussed. Photoresection has a greater range of applications at present but photodynamic therapy has an exciting future potential for cures of early inoperable tumours.

Thermal Effects of Class IV Lasers

Three lasers have been evaluated in thoracic applications: the carbon dioxide, argon and neodymium yttrium aluminium garnet (Nd-YAG) lasers. The carbon dioxide laser is invisible in the far infrared (wavelength 10,600nm) of the electromagnetic spectrum. The beam

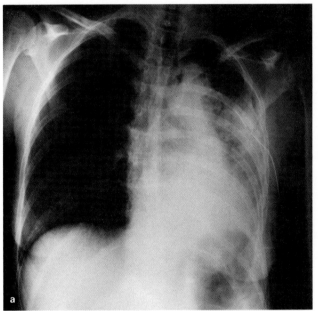

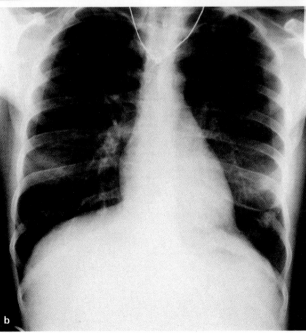

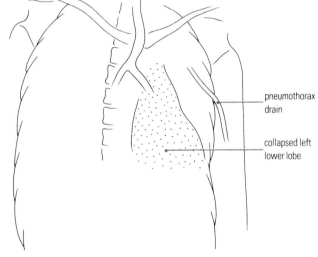

Fig. 6.12 Case study 5. Chest radiographs following treatment for a pneumothorax. In view (a), there is persistent left lower lobe occlusion, but the pneumothorax has been drained. After bronchoscopy and removal of mucus plugs, the lower lobe has re-expanded and the pneumothorax drain has been removed (b).

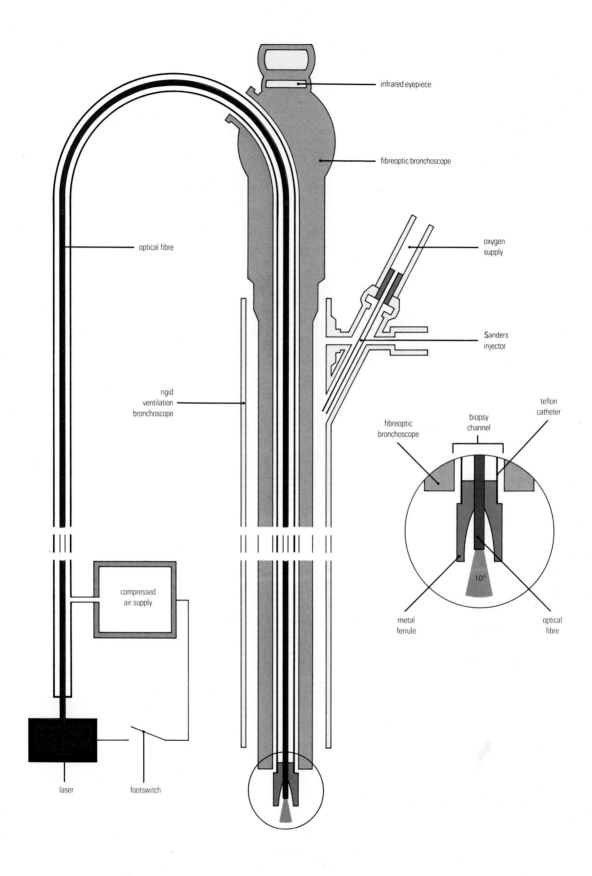

infrared eyepiece

fibreoptic bronchoscope

oxygen supply

Sanders injector

optical fibre

rigid ventilation bronchoscope

fibreoptic bronchoscope

biopsy channel

teflon catheter

10°

metal ferrule

optical fibre

compressed air supply

laser

footswitch

a

6.10

is absorbed into cell water and only penetrates approximately 1mm below the tissue surface. This gives it considerable precision, almost analogous to cutting with a scalpel. Its great disadvantage is that it cannot easily be transmitted through optical fibres. (This is possible with the use of toxic materials in the manufacture of very specialized fibres but is quite impracticable for clinical use.) Consequently, it has only been used in rigid systems. Although it was the first laser used in the bronchial tree during the last decade, it is only effective in the larynx and trachea. As such, it is extensively used in ear, nose and throat (ENT) surgery.

The argon gas laser is visible in the blue-green spectrum (wavelength 514nm) and can be transmitted through optical fibres. It penetrates to approximately 2mm below the tissue surface and has been effectively used in airway surgery with the fibreoptic bronchoscope. However, it is not ideal, because its blue colour results in considerable absorption of energy by blood. It is therefore relatively slow in destroying tumours if there is significant surface bleeding or blood clot on the tumour.

The Nd-YAG is a crystal laser (wavelength 1,060nm, invisible in the near infrared). It is the most suitable laser currently available and is transmissible in fibres of 600μm in diameter. It displays less colour specific absorption which results in quicker tumour destruction than the argon laser. It is absorbed to approximately 6mm below the tissue surface and achieves good haemostasis. The main disadvantage is that its relatively deep penetration produces a destructive rather than a cutting effect and this can obscure anatomical landmarks during treatment. Further discussion will be confined to the Nd-YAG laser.

Nd-YAG System

The main components of airway surgery with the Nd-YAG laser are illustrated in Figure 6.13. The laser beam is focused onto the proximal end of a 600μm quartz or glass fibre which has an outer cladding to reduce emission through its wall. It passes through the fibre by internal reflection. The distal end has no lens

on it and the emerging laser beam has a divergence of approximately ten degrees. This produces a beam which has an effective operating distance in the region of 1cm from the fibre tip, causing relatively little damage at greater distances.

Since the Nd-YAG is invisible, a lower power helium-neon laser is coupled to it producing a red spot with which the Nd-YAG laser can be aimed. An infrared filter is incorporated into the eyepiece of the fibrescope to protect the bronchoscopist's eye from any back reflection through the fibrescope. The fibre is enclosed in a teflon catheter, through which compressed air is blown. This prevents any debris from collecting on the fibre tip should it touch the tumour. Such debris would absorb the energy of the laser and generate temperatures high enough to chip the tip of the fibre and distort the laser beam. Nevertheless, progressive damage does occur to the fibre tip which has to be recut periodically. The fibre tip can deteriorate rapidly and must therefore be kept well clear of the end of the fibrescope, which may otherwise be damaged by scatter of the laser beam from the damaged fibre tip.

The laser is activated by a footswitch which also increases the airflow. The power level at the distal end of the fibre is checked before use by firing it into a power meter to ensure that the laser is functioning correctly and that the fibre is in good condition. Power levels are not critical. The total energy used (measured in joules) is the main determinant of the biological effect. Exposures of 50-70 watts for two seconds duration are usually needed. This is the longest period that can be used without some movement of the target; also longer exposures accelerate fibre tip decay. Lower powers may give effective coagulation.

Laser Treatment under Local Anaesthesia

The laser can be used through the fibreoptic bronchoscope under local anaesthesia. Because of the size of the teflon catheter, most fibreoptic waveguides will only pass through wide-channel bronchoscopes (e.g. Olympus

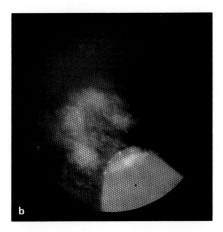
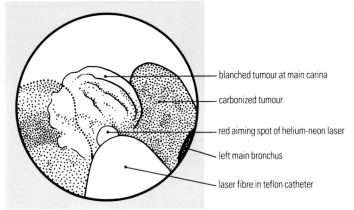

blanched tumour at main carina

carbonized tumour

red aiming spot of helium-neon laser

left main bronchus

laser fibre in teflon catheter

Fig. 6.13 Nd-YAG laser system for bronchoscopic use. The diagram (a – see opposite) shows how the laser beam is focused onto the proximal end of an optical fibre enclosed in a teflon catheter through which compressed air is blown. This is passed through the biopsy channel of a wide-channel fibreoptic bronchoscope. Details of the fibre tip are shown in the inset. The fibrescope is used to aim the laser fibre waveguide. The

fibrescope is best used through a rigid ventilation bronchoscope with a Sanders injector. A footswitch controls the laser and increases gas flow when it is being fired. View (b) shows an endoscopic view of the fibre tip through the fibrescope. A red aiming beam is provided from a coupled helium-neon laser. Blanching and charring resulting from photoresection of a tumour can be seen at the carina.

BFIT10, 2.6mm channel). Treatment rarely causes any pain but smoke provokes coughing. Patients need sedation, with diazepam for example, supplemented with small doses of intravenous diamorphine to suppress

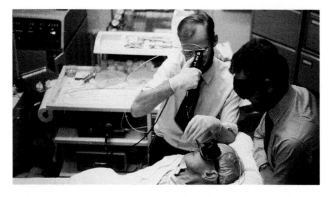

Fig. 6.14 Use of fibreoptic bronchoscope and Nd-YAG laser under local anaesthesia. The fibre waveguide can be seen running from the laser (on the left) into the fibrescope. The assistant wears goggles, but the bronchoscopist is protected by a safety filter in the fibrescope.

coughing. This technique can work well with small obstructions but has important limitations.

Few patients can tolerate prolonged laser photoresection and it is rarely possible to clear the airway fully in a single treatment. Apart from the morbidity of repeated treatments, there is also a danger that any exudation from the treated area could temporarily increase the degree of obstruction. It is impractical to attempt complete vaporization of the tumour since this requires large amounts of energy with considerable smoke generation. Debris must therefore be removed as the tumour is carbonized and this is a slow process with fibreoptic biopsy forceps.

Smoke particles can cloud the fibrescope lenses and the fibrescope may have to be removed for cleaning and then re-introduced through the larynx. The laser will not seal vessels over 2mm in diameter. In patients with severely compromised airways, even small haemorrhages can cause asphyxia. When using only the fibrescope, it is not possible to ventilate the patient while attempting to suck blood out of the airway. Use

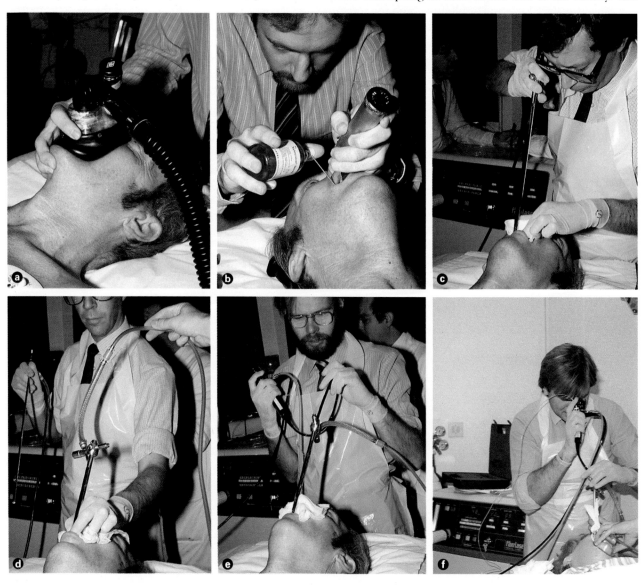

Fig. 6.15 Use of both the fibreoptic and rigid bronchoscope with the Nd-YAG laser under general anaesthesia. The following views are shown: (a) the patient is ventilated by hand with a mask as muscle relaxants start to work; (b) the vocal cords are sprayed with lignocaine; (c) intubation begins with the rigid bronchoscope; (d) the Sanders injector is connected to the rigid bronchoscope; (e) the fibrescope is passed through the rigid bronchoscope; (f) the laser is fired through the fibrescope with the waveguides in the biopsy channel, and a mask attached to a sucker and held by a nurse is used to remove smoke exhaust from the rigid bronchoscope.

of the fibrescope under local anaesthesia is shown in Figure 6.14.

Combined Treatment with Rigid and Fibreoptic Bronchoscopes under General Anaesthesia

This is the method of choice in the majority of cases. After placement of the rigid bronchoscope in the bronchial tree, the fibrescope, with the laser waveguide passed through its suction channel, is passed through the rigid instrument and used to aim the laser beam (Figs 6.15 and 6.16). Rigid forceps can then be used to remove debris more quickly because of their larger jaws (Fig. 6.17). General anaesthesia provides a controlled operative field so that treatment can proceed more rapidly and can usually be completed in a single session, even with large tumours. The fibrescope and laser waveguide can be removed for cleaning as often as required and, in the event of serious haemorrhage, some continued ventilation can be achieved while attempts are made to control it.

Some workers prefer to use the rigid bronchoscope exclusively. The laser waveguide can then be passed through a deflecting device and used with a rigid telescope. The deflecting device is normally used with a 90° telescope to place flexible forceps in the upper lobes. Several purpose-built rigid deflecting devices are now also available for manipulation of laser waveguides. The fibrescope is more versatile, however, in treating more distal lesions and also the upper lobes. Use of the rigid and fibreoptic bronchoscopes together for laser photoresection is an excellent example of the synergism that can be achieved with the two instruments.

General Anaesthetic Technique

General anaesthesia for laser photoresection in the bronchial tree poses considerable problems and should only be undertaken by an experienced anaesthetist. The technique shown in Figure 6.18 together with similar techniques used by other groups, has now been shown to have an excellent safety record, even in patients with

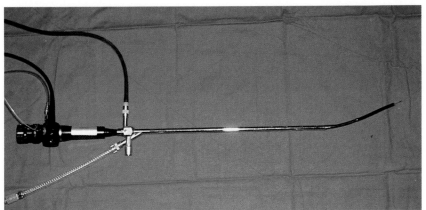

Fig. 6.16 Components used for photoresection. The rigid bronchoscope is shown with inserted fibrescope, laser fibre and attached Sanders injector.

Fig. 6.17 Size comparison of the jaws of the flexible and the larger rigid biopsy forceps. Portions of carbonized tumour debris removed with them are also shown.

General Anaesthetic Protocol for Nd-YAG Laser Photoresection/Photocoagulation

stage of anaesthesia	action
pre-medication on ward	temazepam 20mg orally
on arrival in laser room	pre-oxygenate glycopyrolate 0·4mg i.v.
induction	etomidate 0·1mg/kg for 10 minutes, then continue with 0·01mg/kg/minute when eyelash reflex disappears: alfentanil 10 μg/kg, atracurium 0·6 mg/kg
ventilation	gradually take over ventilation manually and spray cords/trachea with 4% lignocaine pass rigid bronchoscope and ventilate with oxygen and Sanders injector
during treatment	continue alfentanil 0·5-1·0 μg/kg and give boluses of 250-500 μg if blood pressure rises maintain relaxation with atracurium infusion or boluses of 5-10mg
at end of treatment	remove bronchoscope, discontinue drugs, intubate with endotracheal tube and ventilate for 10 minutes with nitrous oxide and oxygen reverse relaxants with neostigmine 2·5mg and atropine 1·2mg i.v.

Fig. 6.18 General anaesthetic protocol for Nd-YAG laser photoresection and photocoagulation. (Developed by Drs C. Nixon and C. Childs.)

very severe obstruction of major airways. The main principles are that a totally intravenous technique is used to avoid any problems of explosion risk with inhalational agents and to facilitate continuous instrumentation through the rigid bronchoscope with an open circuit.

After inducing anaesthesia with etomidate and alfentanil, the patient is paralysed with atracurium and the rigid bronchoscope passed through the larynx. Ventilation is then started with a Sanders injector using one hundred per cent oxygen entraining room air. This gives an oxygen concentration of approximately seventy per cent at the bronchoscope tip, which falls further as a result of mixing in the airways and leakage around the bronchoscope in the larynx. Some workers prefer to use a mixture of fifty per cent oxygen in nitrogen but no evidence of any fire hazard has been found using one hundred per cent oxygen. If severe airway obstruction occurs from haemorrhage during treatment, the high PaO_2 which results from this technique (about 30kPa in most patients) will prevent cerebral damage during quite long apnoeic periods, which may occur if the airway is difficult to clear. This is a valuable safety feature which cannot be provided with spontaneous ventilation and fibreoptic bronchoscopy alone. At the end of treatment the patient is intubated with an endotracheal tube and ventilated with nitrous oxide and oxygen while the atracurium is reversed.

Complications of this technique are rare but include pneumothorax, side-effects of the large cumulative doses of intravenous anaesthetics, such as venous thrombosis at the infusion site, arrhythmias from vagal stimulation and hypoxia. The greatest hazard is accidental underventilation from the reduction in airflow through the rigid bronchoscope caused when the fibrescope is passed through it. In most patients, although there would appear to be little space for gas flow through the rigid bronchoscope, this is not a problem (Fig. 6.19). It may be necessary, however, to remove the fibrescope periodically to allow the anaesthetist to ventilate the patient more efficiently.

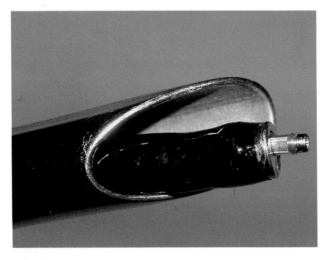

Fig. 6.19 Distal view of fibre tip within fibreoptic bronchoscope. The fibreoptic bronchoscope has been passed through the rigid bronchoscope.

Laser Safety

Class IV lasers will cause permanent retinal damage after even brief exposures, and for this reason they should be installed in a permanent site, such as an endoscopy unit or theatre, which is guarded by warning signs and interlocks which inactivate the laser should anyone attempt to enter the room while it is being fired. In hospitals in the United Kingdom, it is now a Department of Health and Social Security requirement that there is a designated Laser Safety Officer responsible for these arrangements. Endoscopic work is inherently safe because the laser is released within the patient. Accidental firing of the laser while the fibre is outside the bronchoscope is, however, possible and it is therefore recommended that safety goggles be worn. Eyepiece filters in the fibrescope are mandatory. These can be permanently incorporated in most fibrescopes by either manufacturers or distributors.

Medical lasers are complex machines which, at present, have a low level of reliability compared with most medical equipment. They cannot be efficiently and safely used without the assistance of a medical physicist, who is prepared to take a special interest in the equipment and provide local assistance with minor breakdowns and repair of the optical fibre waveguides. It is vital to have several spare fibres which can replace broken ones should they fail during treatment.

Selection of Cases for Laser Therapy

In the palliation of tracheobronchial tumours, the Nd-YAG laser is used to relieve dyspnoea or recurrent haemoptysis in patients who have inoperable tumours and have failed to respond to, or have relapsed after, radiotherapy or chemotherapy. It may also be the only possible treatment in patients who present late with severe stridor, where upper airway obstruction is too severe for the patient to tolerate radiotherapy with the delay in response that this involves. Therefore, its advantages are that it can produce an immediate improvement, has no systemic toxicity and, although not a cure, can be repeated when symptoms recur. Patients should have dyspnoea or haemoptysis as their most troublesome symptoms but the presence of metastatic disease is not a bar to treatment, provided it is not the main cause of symptoms.

The principal limitation of laser photoresection is that it will only help obstruction from tumour within the lumen of the airway. Tumours causing extrinsic compression cannot be treated and attempts to do so, with destruction of the bronchial mucosa and wall, are potentially dangerous. Many patients, however, have a combination of intra- and extraluminal tumour and removal of the intraluminal part alone may still produce a significant improvement. Benefits from photoresection diminish with progression more peripherally because of the branching of the bronchial tree; thus even a small increase in the lumen at a tracheal tumour may produce marked improvement in airflow. Careful bronchoscopic assessment is needed before treatment and, with experience, it is possible to achieve a high response rate and avoid unhelpful treatment for unsuitable patients.

Haemoptysis can usually be controlled, provided the site of bleeding can be clearly identified and is within range of the fibrescope. Haemostasis is simply achieved by blanching the tumour surface at low power. To maintain haemostasis, however, as much tumour tissue as possible should be removed to reduce the volume of tumour vessels from which further bleeding can occur. As with airway obstruction, treatment can be repeated for relapse.

Operative mortality is in the order of one per cent of treatments in experienced hands and is mainly caused by asphyxia from moderate haemorrhage in already severely compromised airways. Resuscitation attempts are likely to be confined to suction and possibly balloon catheter occlusion of a bronchus. The bronchoscopist must be fully aware of this risk and be satisfied that the patient has inoperable disease before attempting treatment with the laser.

The following case studies illustrate the types of patient suitable for treatment. All were treated by combined use of rigid and fibreoptic bronchoscopes under general anaesthesia.

Case Study 6

A sixty-six-year-old man was referred from another hospital where he had presented with a six-week history of rapidly increasing breathlessness which had originally been attributed to asthma. As his stridor became more obvious he was bronchoscoped and tumour was found at the main carina. He was deteriorating rapidly with near asphyxia and emergency laser treatment was preferred to radiotherapy.

At laser treatment, the tumour was found to be causing severe obstruction to the orifices of both main bronchi (Fig. 6.20). The tumour was cut back to its base on the carina and posterior tracheal wall which gave substantial improvement in the airway (Fig. 6.21); pulmonary function tests (Fig. 6.22) showed great improvement. Subsequent investigations showed that the tumour was confined to the base of the trachea and main carina. The patient was therefore referred for elective curative surgery by resection of the carina and lower trachea with anastomosis to the main bronchi.

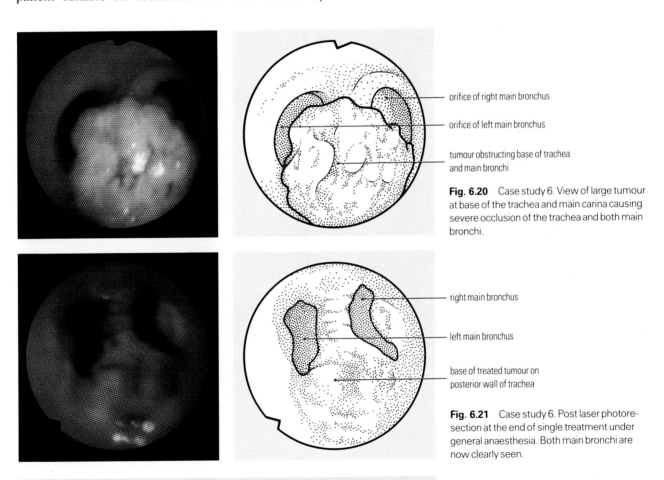

— orifice of right main bronchus

— orifice of left main bronchus

tumour obstructing base of trachea
and main bronchi

Fig. 6.20 Case study 6. View of large tumour at base of the trachea and main carina causing severe occlusion of the trachea and both main bronchi.

— right main bronchus

— left main bronchus

base of treated tumour on
posterior wall of trachea

Fig. 6.21 Case study 6. Post laser photoresection at the end of single treatment under general anaesthesia. Both main bronchi are now clearly seen.

Pulmonary Function Tests

	PEFR (litres/min)	FEV$_1$ (litres)	FVC (litres)
pre laser photoresection	60	0·89	1·02
post-treatment	400	2·59	3·62

Fig. 6.22 Case study 6. Pulmonary function tests before and after laser photoresection treatment for a squamous cell carcinoma at the base of the trachea and main carina.

Case Study 6 illustrates another valuable function of the laser. In patients with life-threatening stridor from tumours, it will not only provide immediate relief but also provide extra time in which patients can be fully staged. In a few patients this can then lead to elective surgery aimed at cure.

Case Study 7

A fifty-eight-year-old man was treated four years previously for a squamous cell tumour of the right main bronchus by right pneumonectomy. He presented after three and one half years with stridor and complaints of profuse sputum which he found difficult to cough up. Bronchoscopy showed multiple tumour deposits in the trachea, some of them occupying over sixty per cent of the tracheal lumen. He was treated with radiotherapy with some improvement but symptoms returned within three months. He was therefore referred for laser photoresection.

During treatment under general anaesthesia he was found to have five tracheal metastases from the original tumour. One of them, at mid-tracheal level, was 2cm long and causing a sixty per cent occlusion. These lesions were all resected at the same treatment session (Fig. 6.23). Although he remained breathless when climbing stairs, he showed considerable improvement in symptoms and was no longer troubled by retention of mucus in the trachea. Pulmonary function also improved (Fig. 6.24). At follow-up bronchoscopy two months later, only one small recurrence of tumour was seen and this was removed at the same diagnostic bronchoscopy using the laser under local anaesthesia. He remains well, eighteen months after his first laser treatment, with periodic elective laser treatment at four-monthly intervals to remove small tumour recurrences. He has no significant symptoms.

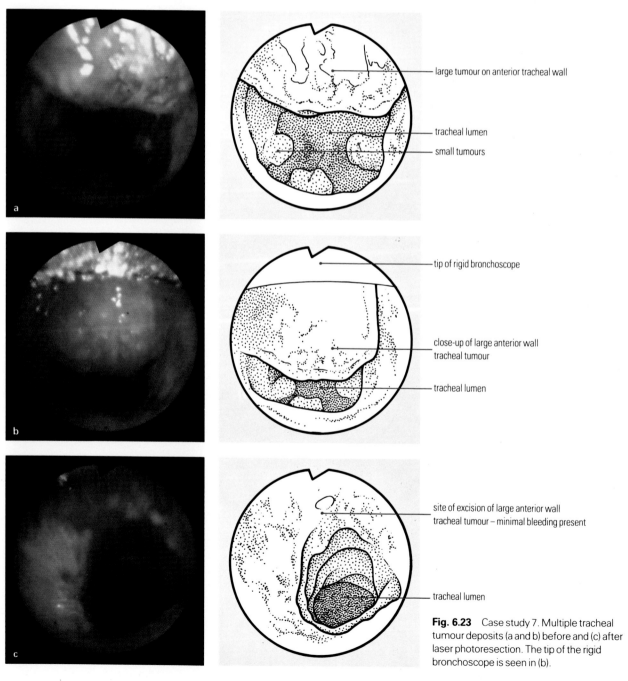

Fig. 6.23 Case study 7. Multiple tracheal tumour deposits (a and b) before and (c) after laser photoresection. The tip of the rigid bronchoscope is seen in (b).

Case Study 8

A twenty-seven-year-old woman underwent hysterectomy for a mesonephric tumour of the cervix at eight years of age. At the age of twenty-five years she developed pulmonary and pleural metastases in the left hemithorax. She was treated with medroxyprogesterone acetate and subsequently cis-platinum. Despite these measures, her pulmonary disease gradually progressed. She was referred for laser photoresection of an endobronchial metastasis which had caused total collapse of the left lung. She had become more breathless as a result of this, but was most distressed by severe haemoptyses and intermittent coughing up of pieces of tumour.

At rigid bronchoscopy under general anaesthesia, a pedunculated tumour was identified in the orifice of the left main bronchus. During resection, it was found to arise more distally from the left lower lobe. It was possible to free completely the upper lobe orifice from tumour but there was clearly a large tumour mass in the lower lobe which also bled very freely (Fig. 6.25). There was little improvement in pulmonary

Pulmonary Function Tests

	PEFR (litres/min)	FEV$_1$ (litres)	FVC (litres)	6 min. walk (metres)
pre laser photoresection	130	0·72	1·69	700
post-treatment	200	1·0	2·51	750

Fig. 6.24 Case study 7. Pulmonary function tests before and after laser photoresection treatment for tracheal metastases.

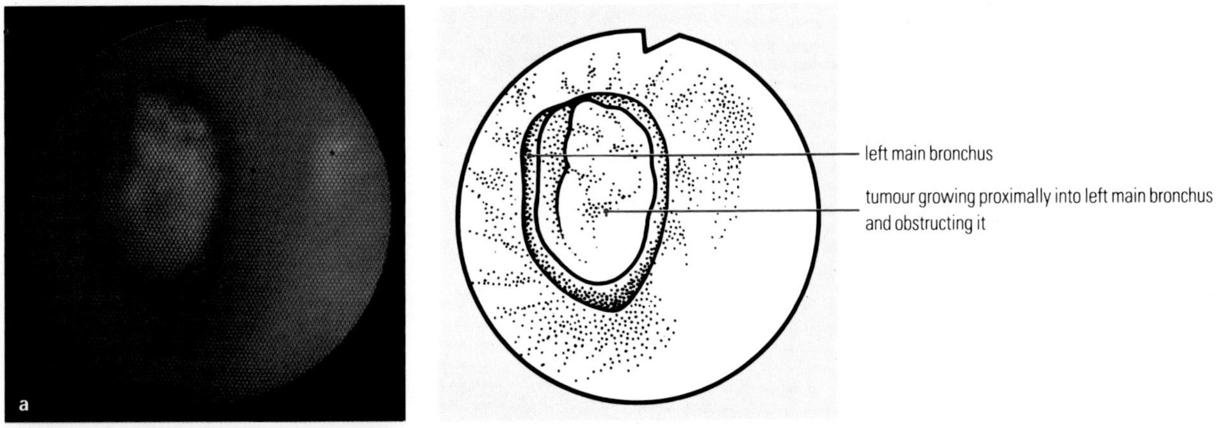

left main bronchus

tumour growing proximally into left main bronchus and obstructing it

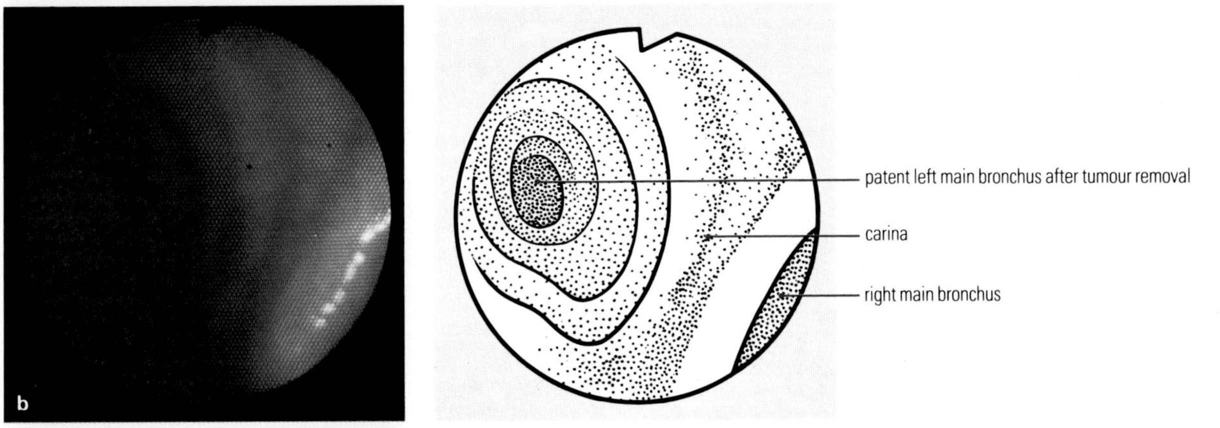

patent left main bronchus after tumour removal

carina

right main bronchus

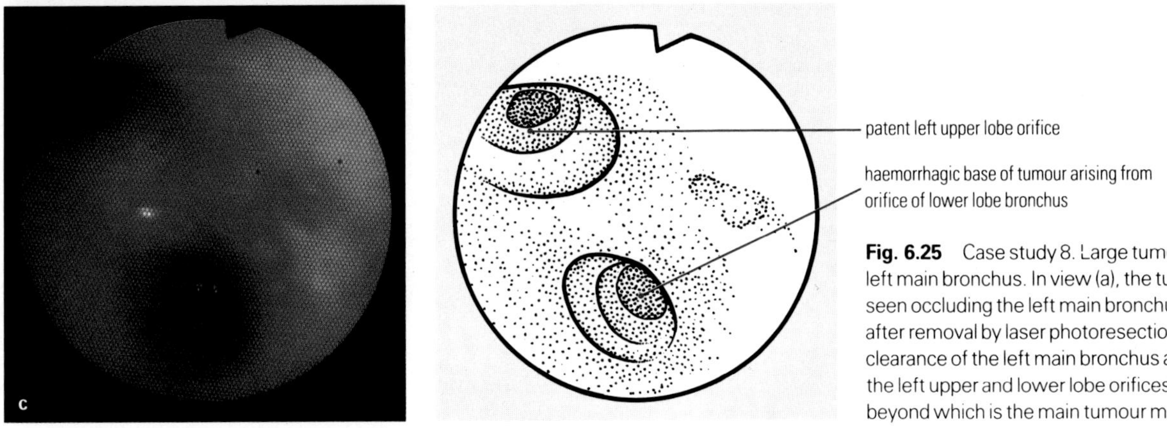

patent left upper lobe orifice

haemorrhagic base of tumour arising from orifice of lower lobe bronchus

Fig. 6.25 Case study 8. Large tumour of the left main bronchus. In view (a), the tumour is seen occluding the left main bronchus. In (b), after removal by laser photoresection, there is clearance of the left main bronchus and in (c) the left upper and lower lobe orifices are seen, beyond which is the main tumour mass.

function since the re-expanded left upper lobe (Fig. 6.26) was invaded by several peripheral metastases. Nevertheless, she gained considerable symptomatic benefit in that her distressing presenting symptoms ceased and her exercise tolerance also improved. She enjoyed good palliation with further treatments for over a year before she died.

As seen in *Case Study 8*, laser photoresection should be considered when complete collapse of a lung or lobe has occurred but success is much less likely than in incom-

plete occlusion of the airway. Three major problems have to be considered when dealing with a collapsed lung: first, it is neither known how great a depth of tumour has to be removed to reach patent airways nor in which direction to cut; secondly, the collapsed lung may be extensively invaded by tumour or damaged by prolonged collapse and may not function even when re-expanded; thirdly, infections are common in re-expanded lungs and can occasionally prove fatal.

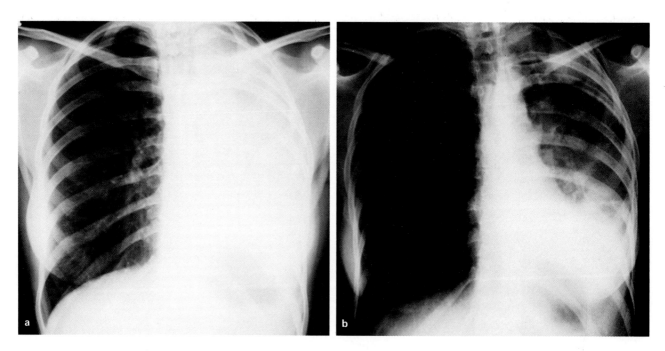

Fig. 6.26 Case study 8. Large tumour of the left main bronchus. (a) chest radiograph before treatment showing complete collapse of the left lung. (b) after treatment there is re-expansion of the upper lobe.

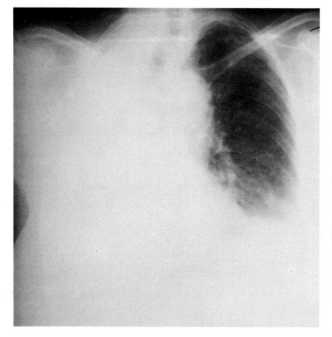

Fig. 6.27 Case study 9. Radiograph showing complete collapse of the right lung due to a squamous cell tumour in the right main bronchus.

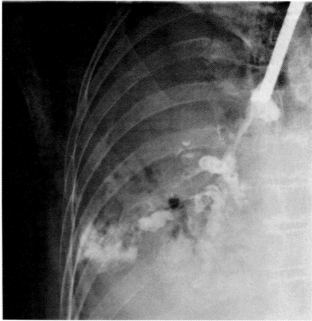

Fig. 6.28 Case study 9. Bronchogram of the middle and lower lobe bronchi in a patient with a squamous cell tumour in the right main bronchus.

Conventional radiography and computed tomography (CT) scans are of no help with the first two problems. Recently, however, improved results have been obtained using bronchography. This technique involves a preliminary fibreoptic bronchoscopy in which an angiography catheter is passed through the fibrescope and then pushed through the tumour; if necessary using a guide wire. Contrast is then injected through the catheter. If the catheter has traversed the tumour to reach the distal bronchi a bronchogram will be obtained; if it has not, the catheter is moved and further contrast injected. A positive bronchogram will indicate that re-expansion is possible and will also show the depth of tumour which has to be explored with the laser. Water-soluble contrast should be used initially to avoid risk of oil embolism should a pulmonary vessel be inadvertently penetrated. This technique is undergoing clinical trials at present. In addition to facilitating successful treatments, it is hoped that a negative bronchogram may be a reliable indicator of cases in which treatment will be ineffective; thus avoiding distress to patients and wasted endoscopy time.

The problem of infection is poorly understood. In laser therapy, infections have only been encountered when re-expanding collapsed lungs. Presumably the inspissated mucus in the collapsed lung provides an ideal culture medium and the prolonged instrumentation necessary for treatment may be contributory. Trap specimens can be taken from the collapsed lung immediately re-expansion occurs, which means that cultures are often available if and when pneumonia occurs. A wide variety of organisms are found and prophylactic broad spectrum antibiotics would therefore appear to be indicated.

Case Study 9

A seventy-six-year-old Greek woman presented in Greece with a two-month history of haemoptysis and progressive breathlessness. Bronchoscopy revealed a squamous cell tumour in the right main bronchus with near total occlusion. She was given cyclophosphamide and subsequently sent to London for radiotherapy which proved unhelpful. She was referred for laser therapy two months after the initial diagnosis, by which time the right lung had completely collapsed (Fig. 6.27).

At fibreoptic bronchoscopy, a catheter was successfully passed through the tumour and a bronchogram obtained of the middle and lower lobe bronchi (Fig. 6.28). Laser resection was therefore attempted under general anaesthesia. The tumour was cleared from the intermediate bronchus and found to have originated from the upper lobe which could not be opened up. Postoperative radiographs showed re-expansion of the middle and lower lobes (Fig. 6.29). There was significant improvement in symptoms and lung function tests (Fig. 6.30).

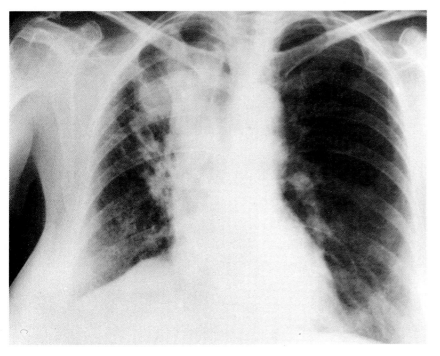

Fig. 6.29 Case study 9. Postoperative radiograph showing re-expansion of the middle and lower lobes.

Pulmonary Function Tests

	PEFR (litres/min)	FEV₁ (litres)	FVC (litres)	6 min. walk (metres)
pre laser photoresection	205	1·1	1·43	100
post-treatment	265	1·42	1·87	375

Fig. 6.30 Case study 9. Pulmonary function tests before and after laser photoresection of a squamous cell tumour in the right main bronchus.

Case Study 10

A seventy-year-old-woman presented seven years previously with breathlessness from complete collapse of the left lung. Bronchoscopy showed the presence of a tumour at the orifice of the left main bronchus and subsequent histology confirmed it as an adeno-carcinoma. No definitive treatment was given at that time. She surprised her doctors by surviving with no worsening of her symptoms and a repeat bronchoscopy and biopsy was reported as a carcinoid tumour. She remained reasonably well until, after five years, she developed troublesome haemoptyses and was referred for laser photocoagulation.

She was initially treated under local anaesthesia when haemorrhagic tumour (Fig. 6.31) was seen arising from the orifice of the left main bronchus and encroaching upon the orifice of the right bronchus and distal trachea. The surface of the tumour was easily coagulated and her haemoptyses ceased. Symptoms were well controlled for one year after which she was again treated for recurrence of haemoptyses and mild stridor. In addition, more serious obstruction at the carina was found; thus a more thorough clearance was performed under general anaesthesia. Some improvement in pulmonary function was seen (Fig. 6.32). No attempt was ever made to re-expand the left lung as it was known to have been collapsed for five years on first presentation.

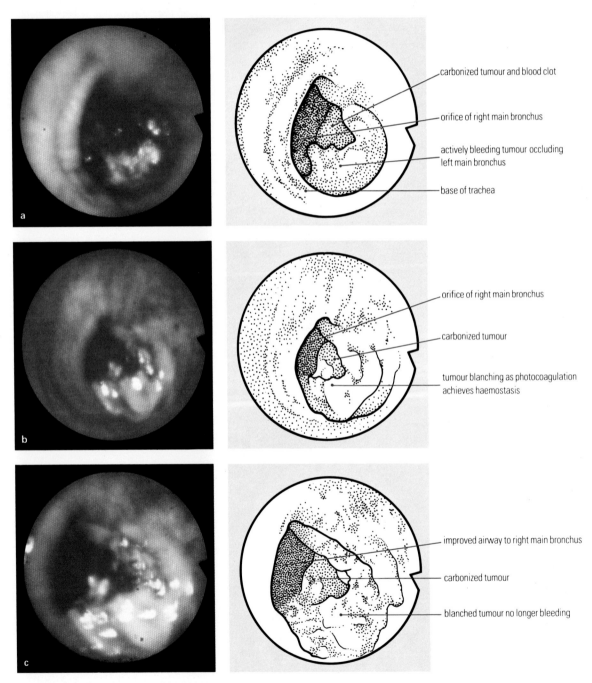

Fig. 6.31 Case study 10. A carcinoid tumour arising from the orifice of the left main bronchus, causing some reduction in tracheal lumen. In view (a), it is actively bleeding. With laser photocoagulation, blanching occurs and haemostasis is achieved. Some carbonization is also seen.

She survived for three years with repeated laser photoresection, at approximately six-monthly intervals to protect the remaining airway to the right lung. She died suddenly from myocardial infarction, but at postmortem she was also found to have extensive local spread of her carcinoid tumour.

Case Study 10 illustrates that haemoptyses can be well controlled, provided the bleeding site can be clearly identified. Similar results have been achieved with malignant tumours causing haemoptysis (Fig. 6.33). Carcinoid and other slow-growing, low-malignancy tumours that are inoperable are ideal for laser photoresection, because of the longer periods of remission of symptoms that can be achieved between treatments. Moreover, death often results from airway obstruction in bronchial adenomas, thus laser photoresection may improve longevity in these cases.

Pulmonary Function Tests

	PEFR (litres/min)	FEV$_1$ (litres)	FVC (litres)	6 min. walk (metres)
pre laser photoresection	60	0·79	1·36	100
post-treatment	137	0·99	1·39	200

Fig. 6.32 Case study 10. Pulmonary function tests before and after photoresection of a carcinoid tumour arising from the orifice of the left main bronchus.

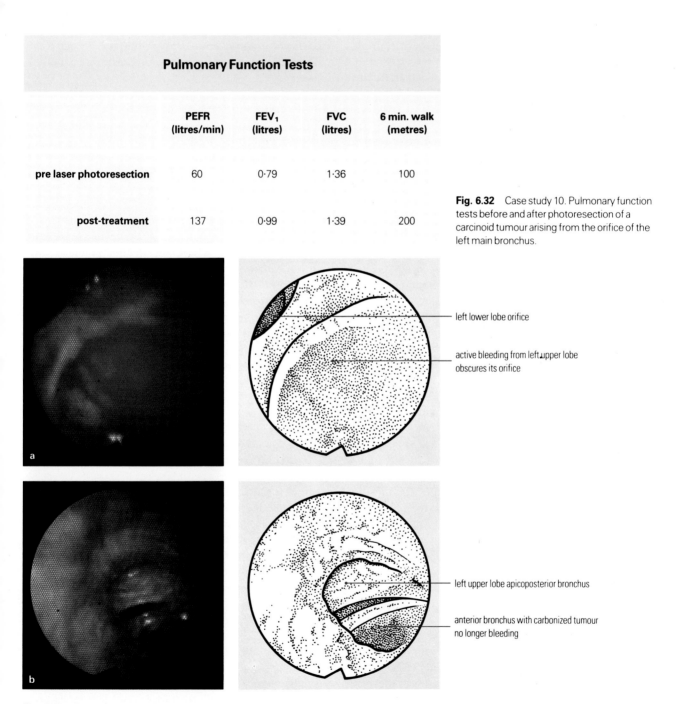

left lower lobe orifice

active bleeding from left upper lobe obscures its orifice

left upper lobe apicoposterior bronchus

anterior bronchus with carbonized tumour no longer bleeding

Fig. 6.33 Control of haemorrhage from an upper lobe tumour. This seventy-five-year-old patient had troublesome haemoptysis from squamous cell carcinoma of the left upper lobe bronchus, which had recurred after previous maximal doses of radiotherapy. The bleeding site could only be reached with the fibrescope, but this was placed within the rigid bronchoscope to achieve better operative control of bleeding. Active bleeding (a) is seen to have come from a distal tumour in the upper lobe. The anterior segment (b) with its carbonized surface is shown after treatment.

Photodynamic Therapy with a Haematoporphyrin Derivative (HPD)

This is an alternative palliative method which can be used on the same types of tumour as the Nd-YAG laser. HPD is an acid derivative of haematoporphyrin, now also available in a purer form, dihaematoporphyrin ether (DHE). Prior to bronchoscopy, it is given by slow intravenous injection in a dose of 3-5mg/kg body weight. Bronchoscopy is then performed seventy-two hours later, normally with local anaesthesia, and an optical fibre is passed through the suction channel to irradiate the pre-treated tumour with red light with a wavelength of 630nm. This is an absorption peak for HPD and also, very conveniently, a wavelength at which light has good transmission through the tissues.

Light of this wavelength can be produced from an argon dye laser (wavelength 514nm) by passing it through a jet of rhodamine B dye which changes the wavelength to 630nm. Alternatively, the gold vapour laser produces red light at 628nm. HPD and DHE are retained by tumour cells and, on excitation with red light, they exert a cytotoxic effect by formation of singlet oxygen. During the next two to three days, the tumour forms a slough; much of this effect is believed to result from damage to tumour vasculature and cell death occurs between one and ten hours after treatment.

The advantages of photodynamic therapy are that only low laser energies are required; approximately $200J/cm^2$ compared with about $4,000J/cm^2$ in photoresection and vaporization with the Nd-YAG laser. These are delivered at powers of about one watt over six to eight minutes. Thus there is little damage to the optical fibre which does not need a compressed air jet for protection. Photodynamic therapy can also penetrate more deeply and may be more valuable in re-expansion of collapsed lungs.

The most important use for photodynamic therapy may be in the attempted cure of early but inoperable tumours. Some Japanese workers claim cures in a small number of patients at follow-up of two to three years.

The active component of HPD, DHE, can be used in lower doses of 2-3mg/kg body weight which, it is hoped, will result in less photosensitization of the skin. This side-effect persists for approximately one month after treatment, since there is some retention of HPD in the skin and other normal tissues with relatively fast cell turnover. However, this photosensitivity is not normally a serious problem if direct sunlight is avoided and barrier creams are used.

There are important disadvantages, however. The tumour slough is very viscid and may cause further, potentially dangerous, airway obstruction. Follow-up bronchoscopy to remove it after two to three days is mandatory; thus palliation cannot be safely given with a single bronchoscopy as is possible when using the Nd-YAG laser. There is no immediate visible effect at the time of treatment and the bronchoscopist must judge the correct dose to give. The Nd-YAG laser seals small blood vessels and promotes some fibrotic response from sur-rounding tissues; in contrast, photodynamic therapy does not appear to have these effects and a few patients have died from haemorrhage one to two days following treatment.

There are currently no adequate comparative studies of the Nd-YAG laser and photodynamic therapy and treatment is, to some extent, a matter of personal preference. The special facilities needed for the therapeutic use of lasers means that few workers find it practicable to use both systems and have therefore chosen one or the other. Ideally, however, one might use both; the Nd-YAG laser being used first to establish a satisfactory airway, followed by photodynamic therapy to treat a greater mass of tumour and prolong remission.

HPD will also fluoresce on exposure to light at 405nm from a krypton-ion laser and this technique has been used in the localization of *in situ* carcinoma for more accurate biopsies.

USE OF LASERS IN NON-MALIGNANT CONDITIONS

This field is restricted to the Nd-YAG laser and indications are less well established than in tumours. Some of these will be considered.

Tracheal Stenosis

Laser photoresection can be used for the 'diaphragm' type of stenosis which can be easily cut away to re-establish the airway. It should not, however, be used in the 'bottleneck' type where the deeper layers of the tracheal wall are involved (Fig. 6.34), since this may result in immediate perforation. There is some scarring and contraction when a laser lesion heals and those studies which report use of the laser for this type suggest that re-stenosis is likely to occur. It should not be used if conventional surgical resection of the stenosis is possible. For 'bottleneck' strictures which are inoperable, a Montgomery T tube stent should be considered. This is well illustrated in *Case Study 11*.

Case Study 11

A forty-one-year-old man who fell off a cliff while on holiday abroad, sustained a head injury and required mechanical ventilation. A poorly placed tracheostomy tube caused a 5cm long tracheal stricture in the upper trachea. Upon his return home, this was treated by resection and re-anastomosis, but subsequently the anastomosis re-stenosed. This was treated with dilatations using the rigid bronchoscope, but re-stenosis rapidly recurred and he became confined to hospital through needing repeat dilatation every two weeks. It was not thought that further trachea could be resected.

He was therefore referred for laser resection on the grounds that conventional surgery was not possible. As seen in Figure 6.35 the stenosis was narrowed to approximately 3mm and after laser resection of the edge of the 'diaphragm' it could just be passed by the fibrescope. Symptoms and lung function tests (Fig. 6.36) improved but he relapsed with re-stenosis after one month. The laser had therefore only marginally improved upon the results of dilatation. The patient was therefore treated with a Montgomery T tube stent which was tolerated quite well.

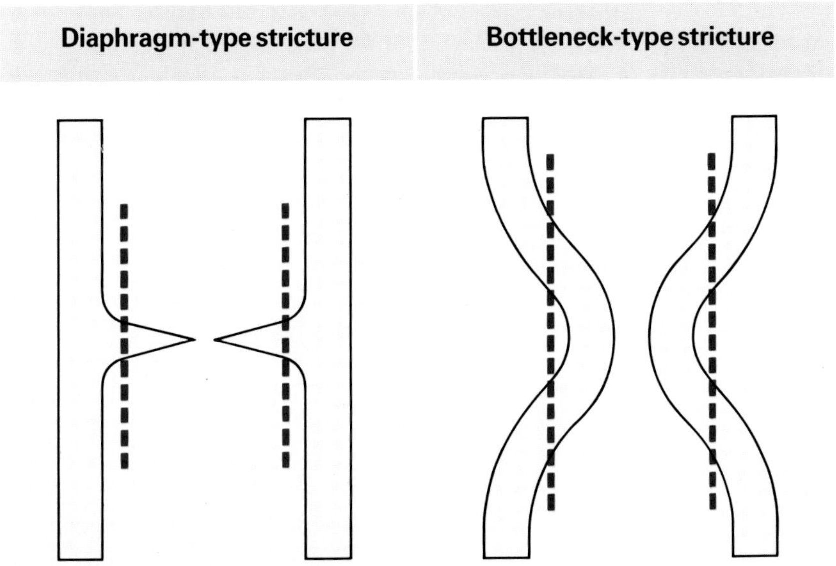

Diaphragm-type stricture	Bottleneck-type stricture

Fig. 6.34 Laser photoresection of tracheal stenosis. The diaphragm type of stenosis may be amenable to resection as indicated by the dotted lines. The bottleneck type, with contraction of deeper layers of tracheal wall, cannot be safely treated since this would result in perforation.

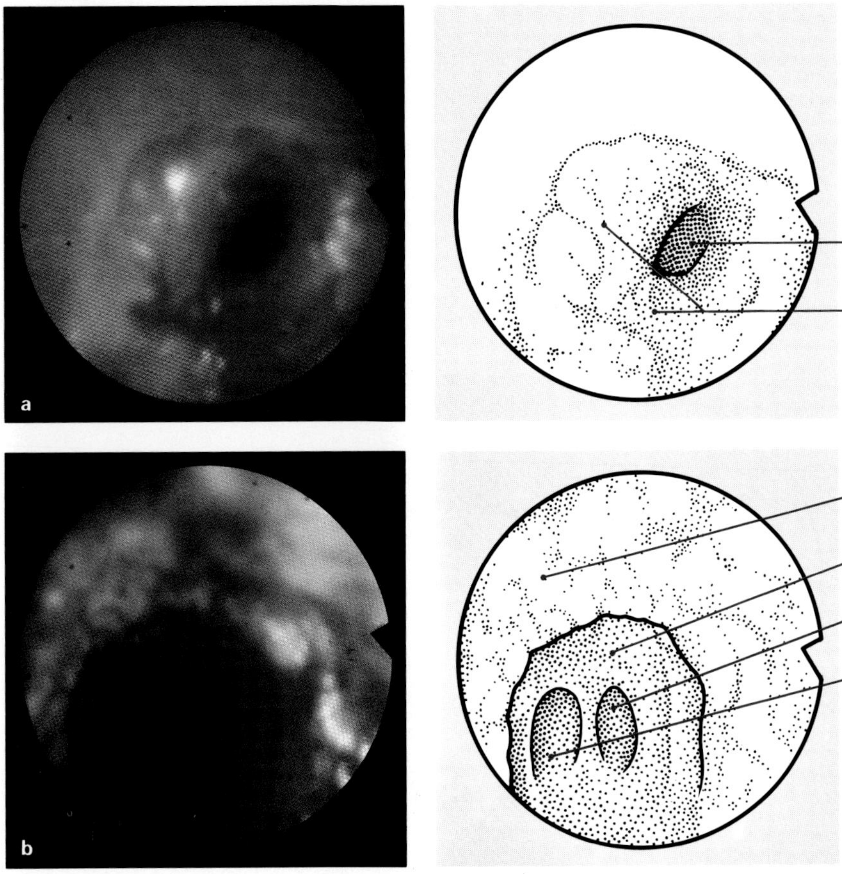

narrow slit of patent airway

'diaphragm' of mucosa stenosing trachea

carbonized edge of stenosis

wider airway achieved through stenosis

right main bronchus

left main bronchus

Fig. 6.35 Case study 11. View (a) is a bronchoscopic view of the tracheal stricture from above. (b) shows the stricture after laser photoresection. The distal trachea and carina can now be seen and the lumen is sufficiently wide to pass the fibrescope through.

Pulmonary Function Tests

	PEFR (litres/min)	FEV₁ (litres)	FVC (litres)
pre laser photoresection	130	1·55	3·3
post-treatment	270	3·08	3·36

Fig. 6.36 Case study 11. Pulmonary function tests before and after laser photoresection for a tracheal stricture in the upper trachea.

Chronic Inflammatory Conditions

A few rare conditions which involve chronic inflammatory processes that cause obstruction of proximal airways and are refractory to other treatments have been described as case studies *(see Chapter 4)*. It is difficult to define the role of the laser for these because their rarity has prevented any proper clinical trials to date. This group includes amyloidosis (of the tracheobronchial type), sarcoidosis (tracheobronchial as opposed to the more common peripheral airway and parenchymal disease) and rhinoscleroma.

Case Study 12

A thirty-eight-year-old woman initially presented with cough and haemoptysis and bronchoscopy revealed a haemorrhagic mass in the right main bronchus and carina. Bronchoscopy was difficult because of the severity of the haemorrhage. The initial histological diagnosis was squamous cell carcinoma and she was treated with radiotherapy with some initial benefit. Her haemoptyses recurred, however, and repeat biopsies were found to show marked squamous metaplasia

overlying nodules of congophilic material which was shown to be primary amyloid tissue (AL type protein fibrils).

Rigid bronchoscopy was then attempted with use of rigid forceps to remove these lesions, but this had to be abandoned because of haemorrhage. She was therefore referred for laser photocoagulation, which was performed under general anaesthesia with both the rigid bronchoscope and fibrescope. The mucosa was studded throughout with amyloid deposits which bled easily. Appearances were most pronounced in the right bronchial tree and this region of the lung alone was treated. Haemoptyses resolved immediately, but approximately one week after treatment she had to be readmitted with collapse of the right upper lobe. This was complicated by infection and did not resolve with physiotherapy and antibiotics. At fibreoptic bronchoscopy, viscid exudate was removed from the upper lobe bronchus, after which she made a good recovery. She has had no significant haemoptyses for two years without further laser treatment. She has, however, had a few episodes of chest infections which appear to be facilitated by obstruction of smaller airways by amyloidosis. These airways are beyond the range of the fibrescope.

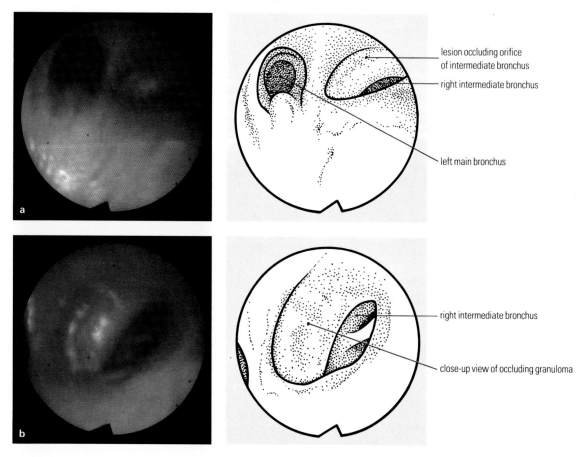

lesion occluding orifice
of intermediate bronchus

right intermediate bronchus

left main bronchus

right intermediate bronchus

close-up view of occluding granuloma

Fig. 6.37   Case study 13. Granulomatous tissue is shown partially occluding the orifice of the intermediate bronchus at the suture line of previous sleeve resection of squamous cell tumour of the right upper lobe; (a) normal and (b) close-up views.

Two other patients referred with amyloidosis were unsuitable for treatment because it was clear that the amyloid tissue was causing extrinsic compression deep to the mucosa. As seen in *Case Study 12*, the laser may provoke a brisk inflammatory reaction in the early postoperative phase and this could severely aggravate the airway obstruction. As these other cases had critical airway narrowing, laser treatment was probably an unacceptable risk. Regrettably, the majority of bronchial amyloid cases involve lesions deep to the mucosa.

Granulomata and Scars on Suture Lines

The laser may have some place for removing granulomata and scars on suture lines if they are genuinely causing symptoms and cannot be removed by conventional means. Lesions which have been subjected to laser therapy will eventually result in some fibrosis and contraction and removing such lesions solely to make the bronchial tree look 'tidy' should be avoided.

Case Study 13

A sixty-seven-year-old man had a sleeve resection for a squamous cell carcinoma of the right upper lobe at the age of fifty-five years. One year later, he was noted to have a granuloma at the suture line with the intermediate bronchus. Ten years later he started a series of recurrent attacks of fever with purulent sputum. Fibreoptic bronchoscopy revealed a substantial flap of scar tissue over the medial side of the intermediate bronchus (Fig. 6.37), which appeared to be causing sputum retention and was the presumed cause of his symptoms. Basal crackles suggestive of bronchiectasis were also noted. The patient's FEV_1 was only 1.1 litres; thus he was referred for laser photoresection of this lesion.

Laser photoresection was carried out successfully (Fig. 6.38) but treatment apparently provoked a further exacerbation of his fever. CT scanning confirmed the presence of lower lobe bronchiectasis. After being taught postural drainage exercises, symptoms improved markedly with only two further attacks over two years follow-up.

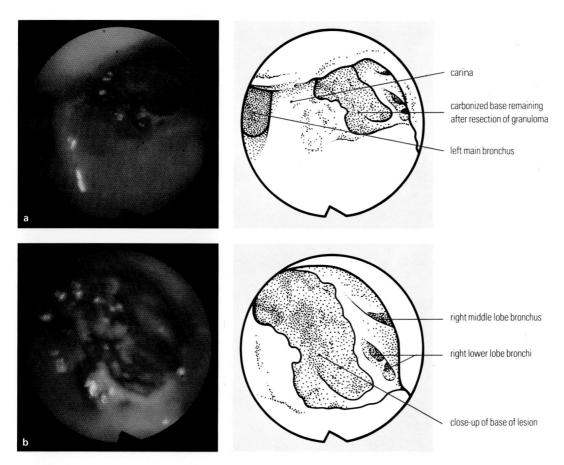

Fig. 6.38 Case study 13. (a) normal and (b) close-up views show the carbonized residue after laser photoresection of suture line granuloma. The lower lobe bronchi are now more clearly visible.

6.25

Pulmonary function tests improved over the next four months (Fig. 6.39). Repeat fibreoptic bronchoscopy at six months showed satisfactory healing of the treated area with no re-stenosis. How much of his improvement was attributable to the physiotherapy and postural drainage is not known. Also it is not known, if laser photoresection had been available when the granulomata were first seen, whether this would have prevented him developing bronchiectasis and his subsequent symptoms.

BRONCHOSCOPIC TECHNIQUES FOR RADIOTHERAPY

An alternative to laser therapy of tumours is interstitial implantation of radionuclides by bronchoscopy. This technique has been available since 1955 and can be performed with rigid or fibreoptic bronchoscopes, although most workers use rigid instruments. In a recent study with radioactive gold grains, Law and colleagues suggested that it is as effective as the laser. In their series of eighteen cases, it appeared most valuable in re-expansion of collapsed lungs and less valuable in tracheal obstruction; somewhat in contrast to experience with the Nd-YAG laser.

This localized radiotherapy can be repeated if required in addition to maximal doses of external radiation given previously. Its disadvantages are that patients must be kept in hospital behind lead screens for a few days and their sputum must be screened for radioactivity from any grains that may be coughed out. Their short half-life makes them relatively harmless after approximately one week. Gold grains are expensive and there is no financial advantage over the laser. They have to be ordered and prepared and, as with other forms of radiotherapy, tumour regression takes time. The laser clearly has advantages for emergency treatment of stridor.

No comparative studies of laser versus gold grains are available. Some interesting pilot studies are now being described of brachytherapy in which the Nd-YAG laser is used to create a cavity in the tumour into which interstitial radionuclides are then implanted.

Pulmonary Function Tests

	PEFR (litres/min)	FEV$_1$ (litres)	FVC (litres)	6 min. walk (metres)
pre laser photoresection	155	1·1	2·5	348
post treatment and postural drainage	180	1·54	3·4	435

Fig. 6.39   Case study 13. Pulmonary function tests before and after photoresection to remove a flap of scar tissue which was presumed to be causing sputum retention.

REFERENCES

Case records of the Massachusetts General Hospital: Case 48 (1983). New England Journal of Medicine, 309, 1374-1381.

Dhillon DP, Collins JV (1984). Current status of fibreoptic bronchoscopy. Postgraduate Medical Journal, 60, 213-217.

Hayata Y, Kato H, Konaka C, Ono J, Takizawa N (1982). Hematoporphyrin derivative and laser photoradiation in the treatment of lung cancer. Chest, 81, 269-277.

Hetzel MR, Nixon C, Edmondstone WM, Mitchell DM, Millard FJC, Nanson EM, Woodcock AA, Bridges CE, Humberstone AM (1985). Laser therapy in 100 tracheobronchial tumours. Thorax, 40, 341-345.

Law MR, Henk JM, Goldstraw P, Hodson ME (1985). Bronchoscopic implantation of radioactive gold grains into endobronchial carcinomas. British Journal of Diseases of the Chest, 79, 147-151.

Lillington GA, Ruhl RA, Pierce TH, Gorin AB (1976). Removal of endobronchial foreign body by fiberoptic bronchoscopy. American Review of Respiratory Disease, 113, 387-391.

Rogers SN, Benumof JL (1983). New and easy techniques for fiberoptic endoscopy-aided tracheal intubation. Anesthesiology, 59, 579-582.

Toty L, Personne C, Colchen A, Vourc'h G (1981). Bronchoscopic management of tracheal lesions using the neodymium yttrium aluminium garnet laser. Thorax, 36, 175-178.

Index